BEING AND OWNING

Being and Owning

The Body, Bodily Material, and the Law

JESSE WALL
Lecturer, Faculty of Law, University of Otago

OXFORD
UNIVERSITY PRESS

OXFORD
UNIVERSITY PRESS

Great Clarendon Street, Oxford, OX2 6DP,
United Kingdom

Oxford University Press is a department of the University of Oxford.
It furthers the University's objective of excellence in research, scholarship,
and education by publishing worldwide. Oxford is a registered trade mark of
Oxford University Press in the UK and in certain other countries

First Edition published in 2015

Published in the United States of America by Oxford University Press
198 Madison Avenue, New York, NY 10016, United States of America

British Library Cataloguing in Publication Data
Data available

Library of Congress Control Number: 2015931223

ISBN 978–0–19–872798–9

To Laura

Acknowledgements

I would like to acknowledge Charles Foster, Joshua Getzler, Imogen Goold, Sarah Green, Jane Kaye, Graeme Laurie, and Ben McFarlane, who all sharpened my thinking by providing comments on earlier versions of this work.

Laura Fraser, Amanda Gilbertson, Marcelo Rodriguez Ferrere, Daniel Vujcich, and Tim Wilson have also helped me clarify my thoughts and expression, and working with John Lidwell-Durnin helped me grasp aspects of phenomenology.

I would also like to acknowledge the support I received from The Rhodes Trust, Pembroke College (Oxford), and Merton College which enabled me to pursue this research. In particular, I am grateful to have been mentored by Mindy Chen-Wishart, Ariel Ezrachi, Jennifer Payne, and Rebecca Williams.

I reserve special mention for Jonathan Herring, whose enthusiasm and insight sustained this project.

Summary Contents

Detailed Contents

Table of Cases

Australia

Canada

The United Kingdom

United States of America

Table of Legislation

Introduction

I. The Property of No One

Common law systems have inherited from the seventeenth century the rule that there is 'no property in the human body'. This rule can be traced back as far as Coke's *Institutes of the Laws of England* in 1644. The rule, as it was stated by Coke, required that: 'the buriall of the *Cadaver* ... is *nullius in bonis* [the property of no one], and belongs to Ecclesiastical cognizance'.[1] As Peter Skegg explains, at that time 'corpses were normally buried in consecrated ground. Once there, they were protected by ecclesiastical law'.[2] The interpretation of the rule was that the 'cognizance' of ecclesiastical courts was at the exclusion of jurisdiction of the civil courts. This implication was not a necessary one. Monuments on consecrated ground, for instance, were both subject to the ecclesiastical jurisdiction and protected by civil actions.[3]

It was not until 1882 that the 'no property rule' was affirmed in a recorded English judgment.[4] In *Williams v Williams*, the deceased had provided in his will that his body be 'given' to his 'friend Miss Eliza Williams, to be dealt with by her in such a manner as [he had] directed ... in a private letter'.[5] The will also directed the executors of his estate to reimburse Miss Williams the expenses incurred from carrying out the instructions. The executors refused to pay Miss Williams the £321 of expenses that she incurred and Miss Williams brought an action against the executors. Her claim was unsuccessful. The

[1] E Coke, *The Third Part of the Institutes of the Laws of England: Concerning High Treason, and Other Pleas of the Crown, and Criminal Causes* (London: 1644) 203; PDG Skegg, 'Human Corpses, Medical Specimens and the Law of Property' (1975) 4 *Anglo-Am L Rev* 412; PDG Skegg, 'The "No Property" Rule and Rights Relating to Dead Bodies' (1997) 5 *Tort L Rev* 222, 222–9.

[2] PDG Skegg, 'Human Corpses, Medical Specimens and the Law of Property' (n 1) 412.

[3] PDG Skegg, 'Human Corpses, Medical Specimens and the Law of Property' (n 1) 412.

[4] *Williams v Williams* (1882) 20 Ch D 659; PDG Skegg, 'Human Corpses, Medical Specimens and the Law of Property' (n 1) 415.

[5] PDG Skegg, 'Human Corpses, Medical Specimens and the Law of Property' (n 1) 415.

Court of Chancery held that since *there is no property in the dead body of a human being*, the deceased was unable to dispose of, or make directions as to, his dead body through his will.

Over the three and a half centuries since Coke's *Institutes*, several claims based on property rights in the human body and its parts have failed on the basis that the human body is not the subject of property.[6] Yet, recently courts have developed a series of exceptions to the 'no property rule'. For example, recent decisions have recognized, in contrast to the decision in *Williams v Williams*, that a deceased person is able to dispose of his stored semen by will,[7] that bodily material can be directed through the application of intestacy law as items of property,[8] and that a widow may obtain the right to possess the body and bodily material of the deceased as items of property.[9]

These recent exceptions to the 'no property rule' reflect the fact that the human body is no longer always destined for consecrated ground. The body and bodily material can be retained, stored, and used for a wide range of purposes. Items of bodily material have become valuable things that people claim the right to possess, use, and control. What has so far remained constant is the way in which we are embodied; our bodies enable our existence, our experiences and our engagement with the world. It is in this way that our bodies have always been valuable things. The use and storage of bodily material is a context where these two values – one constant and one newly emerging – converge.

The current law is therefore evenly poised between a general rule, which states that there is no property in the human body or its parts, and a series of expanding exceptions to the general 'no property rule'. The application of the 'no property rule' leaves a void in the law by only negating the recognition of property rights in bodily material without positing an alternative right in bodily material. Moreover, the expanding exceptions to the general rule leave the scope of the 'no property rule' uncertain. As a result, the legal status of bodily material is undetermined and unclear.

[6] *Dobson and another v North Tyneside Health Authority and another* [1996] 4 All ER 474; *Re: Organ Retention Group Litigation* [2005] QB 506; *Greenberg v Miami Children's Hospital* 264 F Supp 2d 1064, 1066 (SD Fla 2003); *Moore v Regents of the University of California* 793 P 2d 479 (Cal 1990); *Washington University v Catalona* 437 F Supp 2d 985 (ED Mo 2006).

[7] *Hecht v Superior Court of Los Angeles County* 20 Cal Rptr 2d 275 (Cal Ct App 1993).

[8] *Bazley v Wesley Monash IVF Pty Ltd* [2010] QSC 118.

[9] *Jocelyn Edwards; Re the Estate of the Late Mark Edwards* [2011] NSWSC 478.

II. Three Main Inquiries

The aim of this book is to determine the appropriate legal status of bodily material. In doing so, I provide a way of filling the void that is left by the 'no property rule'. The analysis is shaped by three main inquiries or themes that develop throughout the chapters. The first inquiry concerns the distinction between ownership as a functional relationship between a person and thing, and the legal relationships between a rights-holder and duty-bearer that are constructed to protect ownership relationships. According to this distinction, property rights have two dimensions. Property rights ground the ability for a person to use or control an object or resource whilst also providing a particular configuration of legal rights and duties that protect the relationship of use or control over the object or resource. The 'no property rule' negates both these dimensions. It significantly limits the ability for a person to assert the right to possess, use, or control an item of bodily material, and it prevents property law from protecting whatever interest a person may nonetheless have in the bodily material.

To the extent that this book aims to fill some of the void left by the 'no property rule', we can anticipate that such a task is twofold. First, it requires us to identify when, and under what circumstances, a person ought to be able to possess, use, control, transfer, or sell, bodily material. Second, it requires us to identify what it means for an 'incident of ownership' or 'entitlement' in an object or resource to give rise to a property right, and then assess whether property law represents the most appropriate branch of law to protect the ownership relationship between a person and an item of bodily material.

The second inquiry concerns an ambiguity in the body. We are all a complex combination of things; genes expressing, cells dividing, neurons firing, and muscles twitching. We are also more than just a complex combination of things. We are things that attract moral attention. The body is nonetheless the location of both a person's thing-ness (physical existence) and the focal point of our duties of respect for that person. The body is ambiguous insofar as it is capable of more than one meaning: it can be both an *object* (a complex combination of things) and can be the *subject* (that which attracts moral attention). When bodily material is separated from the body the question becomes whether, and to what extent, the bodily material ought to continue to attract the duties of respect that are ordinarily directed towards the person and their body.

If bodily material is only ever a mere object, then the way in which we ought to fill the void left by the 'no property rule' is relatively straight-forward. We ought to allocate and distribute bodily material in the same way that we allocate and distribute other material resources. This may include compulsory redistribution of resources to address social needs and commercial exchanges of resources as an efficient means of resource allocation. Furthermore, if bodily material is only ever a mere object, then the further implication is that the law ought to govern the use and possession of bodily material in the same way it governs the use and possession of other material resources. That is, through the application of property law. This book aims to resist these implications.

The third inquiry concerns a distinction between different sets, or spheres, of value. Whilst we may value a wide range of things, we do not value all things using a single metric or mode of valuation. For instance, we all value our own personal wealth and material goods and we all value our personal relationships and our own reputation. We nonetheless value these things using different modes of valuation. Our bank balance and our friendships clearly belong to different spheres of value. These different spheres of value mirror categories within the law. We enjoy a range of legal rights, such as the right to possess items of property, the right to privacy, and the right to bodily integrity. Our rights with regards to our property enable a set of preferences and choices that another person is able to exercise. The same cannot be said of our rights to privacy or bodily integrity. Such rights protect a set of preferences and choices that can only be exercised by us. I will suggest that there is an underlying distinction that is able to differentiate between spheres of value and categories of legal rights. The distinction suggests that our personal wealth and items of property have a substitutional value that is contingent to any particular person. Our personal relationships, privacy, and bodily integrity, in comparison, have value in their own right and are necessarily associated with us.

In circumstances where bodily material is a mere object, we may value an item of bodily material in the same way that we value our personal wealth or our items of property. In other circumstances, the use and storage of bodily material might engage values that are more akin to the way that we value personal relationships, our privacy, or the use and control over our (living and attached) body. These different modes of valuation, and ultimately different legal categories, follow from the ambiguity in the body and bodily material. What raises concern is when these different values are treated as if they are somehow the same; to reduce the value of personal relationship to a monetary value, or to reduce our bodies to a 'repository' of financially valuable chattels, is to *denigrate* our friendships and our bodies. As we shall see, this concern provides the underlying motivation for maintaining a distinction that is able to differentiate between values, preferences, or choices that are contingently associated with the rights-holder; and values, preferences, or choices that are necessarily associated with the rights-holder.

III. Chapter Synopses

Allow me to briefly state the main arguments in each of the following chapters. In chapter one, I differentiate ownership from property. Ownership is a series or spectrum of functional relationships between a person and an object or resource. To this end, all items of bodily material are capable of being owned since all items of bodily material can, in functional terms, be possessed, used, managed, transferred, sold, and so on. Since each 'incident of ownership' represents a functionally distinct relationship between a person and an object or resource, the task of justifying or explaining why a person ought to be able to exercise an incident of ownership or 'entitlement' in an object or resource will vary depending on the incident or entitlement in question. With regards to items of bodily material, two main types of justifications arise: either entitlements in bodily material can be justified with reference to the 'natural' or 'pre-social' rights of the person or with reference to the state of affairs that follow from the exercise of the entitlements.

Ownership, however, is not a legal concept. The recognition of an incident of ownership or entitlement in bodily material does not necessitate the application of property law. This is because property law represents a particular configuration of rights and duties that protects an ownership relationship. Moreover, there are other branches of law that are able to protect the possession, use, or control of an object or resource. There is a choice, therefore, as to which branch of law ought to govern the possession and use of bodily material.

Chapter two aims to identify when, and under what circumstances, a person ought to be able to possess, control, or use bodily material. Whilst we can provide explanations with relative ease as to why each person has exclusive rights of use and control over their own (living and attached) body, such explanations appear to be unable to extend to bodily material that is separated from the body or extend to the posthumous body and its parts. I assess three possible explanations for why a person ought to retain entitlements in bodily material: the 'prior embodiment principle', the Lockean 'work or skill rule', and the Hegelian 'nexus of dual relations'. I suggest that, from these three explanations, only the Hegelian account is able to justify entitlements in bodily material. The intuitive 'prior embodiment principle' is inapplicable since someone's prior relationship with the body cannot be extended to explain or justify their current assertion of rights in the bodily material. The Lockean 'work or skill rule' is only partially able to bridge the gap between a person and an object or resource by identifying how a person's intentional actions can provide a new attribute in the object or resource. The Hegelian 'nexus of dual relations' is able to fully bridge the gap between subject and object by connecting the object with both a person's intentional actions and an indented state of affairs.

I then consider a fourth explanation for why a person ought to be able to own bodily material. The fourth explanation, following Merleau-Ponty, suggests that our bodies are the site or location of our subjectivity (that which attracts moral duties). Bodily material, in some circumstances, may continue to be the location of our being or subjectivity. The comparison between this explanation and the previous three explanations in this chapter invites an important contrast. The contrast is between two views of the body, as either arbitrarily and contingently associated with the subject, or as necessary and constitutive the subject. How we address the tension between the interests that individuals have in their own bodily material (or the body of a deceased loved-one) and the societal interest in using bodily material for broader healthcare aims is largely informed by whether we view the body as contingently associated or necessarily associated with the person.

Chapter three concerns the sale of bodily material. This chapter advances three main arguments. First, I argue that is not possible to justify the right to profit from the transfer or relinquishment of bodily material with reference to the qualities or attributes of the person. The right to profit, simply put, cannot be a 'natural' or 'pre-social' right. This is because the content of the right to profit is connected to a series of considerations or factors that are external to the rights-holder and the object or resource. Such considerations are of a different order of considerations than those that form the 'natural' or 'pre-social' rights of the person. Second, the right to profit may be justified with reference to the state of affairs that follow from permitting people to exchange bodily material for a financial benefit, such as the increased availability of bodily material for socially valuable enterprises. This justification of the right to profit is valid on the condition that the financial benefit is able to provide reasons for action that motivate decisions to transfer or relinquish bodily material. I will suggest that there are instances where this condition may not be satisfied. Given that we may value some things in their own right, the third argument in chapter three raises the concern that the opportunity to financially benefit from the relinquishment or transfer of bodily material may be the basis of pressure to reduce or denigrate something that has value in its own right into something that has substitutional or equivalent value. The denigration of the bodily material, I suggest, indicates that the ability of a person to determine what physical things constitute their subjectivity may be vitiated.

The second half of the book focuses on questions of law and legal theory. Chapter four identifies two conceptual features of property rights. That is, two sets of the social and moral assumptions that underlie the legal structure of property rights. Property rights, I argue, are exclusive rights. We have an interest in the use of various different things, objects, or resources. This interest is best characterized as an interest in an open-ended set of activities. Property law protects this interest by placing the thing itself as the focal point of the legal relationship between a rights-holder and duty-bearer. The rights-holder then has the ability to exclude all others from the object or resource as a way or protecting and enabling an open-ended set of activities. The right to bodily integrity is conceptually akin to property rights in this regard. The right to bodily integrity is an exclusive right as it enables the rights-holder to exclude all others from their body as way of providing for the rights-holder full authority over an open-ended set of activities with regards to their body.

The second conceptual feature of property rights concerns the relationship between the rights-holder and the preference and choices that the right enables. Property rights are contingent rights; they enable preferences and choices that can exist independently of the rights-holder. We can assess whether a right is a contingent right by applying two criteria. These criteria question whether there are normative consequences (beyond mere allocative consequences) to a change in rights-holder, and question whether another person can stand in the same position as the initial rights-holder with regards to the object or resource. The right to bodily integrity, in contrast to property rights, is a non-contingent right; it protects preferences and choices that can only be exercised by the rights-holder. We can understand why the right to bodily integrity is a non-contingent right either in terms of the close and immediate connection between a person's subjectivity and their body or in terms of the body being the site or location of a person's subjectivity.

Chapter five explores property law as a structure of legal rights and duties. This chapter focuses on five 'structural' or 'doctrinal' features of property rights-based actions, how these structural features follow from the conceptual features identified in the previous chapter, and how these structural features have shaped the way in which the law on the use and storage of bodily material has developed. I suggest that property rights, as conceptually exclusive rights, are rights that are exercisable against an open set of persons, protect the original and derivative dimension of ownership, and impose primary duties of non-interference. This structure of an 'exclusionary boundary' around the object or resource is able to explain why exceptions to the 'no property rule' have been formulated. This exclusionary structure is the same structure applied in legal actions that protect the right to bodily integrity.

As conceptually contingent rights, property rights are presumptively transferable. Moreover, the infringement of property rights are addressed through the imposition of corrective remedial duties that aim to correct or restore the misallocation of entitlements through the award of 'special' damages. The right to bodily integrity, as with other 'personal rights', differs in this regard. Such rights are inalienable and infringements are addressed through the imposition of distributive remedial duties that allocate benefits and burdens in accordance with a criterion of merit through the award of 'general damages'. To the extent that some interests in bodily material may represent a set of preferences and choices that can only be exercised by particular rights-holders, a conceptual and structural inconsistency emerges between property rights and some of the interests that may arise in the use and storage of bodily material.

Chapter six then formulates a recommendation as to how the law ought to govern the use and storage of bodily material. I suggest here that the rights that arise in bodily material are best conceptualized as exclusive rights. However, because of the 'no property rule', the law has so far relied upon a 'governance' rather than an 'exclusion' strategy. The appropriate legal response to the use and storage of bodily material varies, and depends on whether a right is contingently or necessarily associated with the rights-holder. In circumstances where bodily material retains a functional unity with the body of the progenitor (the person from whom the bodily material originates from) or remains as the medium of social experience, the rights that arise in such items of bodily material are exclusive and non-contingent rights. In all other circumstances, rights in bodily material ought to be conceptualized as exclusive and contingent rights.

What complicates the structure of the appropriate legal response to the use and storage of bodily material is that the structure of rights varies between circumstances. An analogy with the right to privacy provides a way forward. The right to privacy is an exclusive but non-contingent right that governs an ambiguous subject matter. Duties of confidentiality with regards to personal information represent a re-adjusted exclusionary boundary that is able to account for the variation in circumstances where information that pertains to an individual does or does not engage a privacy interest. Where a progenitor retains entitlements in bodily material or when family members obtain entitlements in the body of a deceased person, the most appropriate legal structure to protect such entitlements is the same legal structure as the common law right to privacy (facilitated through duties of confidentiality). In all other circumstances, where the bodily material represents a mere material resource, the most appropriate legal structure is the structure of property rights. This dualist approach to the structure of the law follows from an underlying contention throughout the discussion: that although we are a complex combination of things, some things—even when physically separate from us—constitute our being.

IV. Parameters

Finally, allow me to clarify the parameters of this overall inquiry. The inquiry is concerned with the law that governs 'bodily material'. This includes any part of the human body, from the smallest cell to the largest organ, that has been detached or extracted from the body, as well as the body and bodily material of deceased persons. My focus will be on the use and storage of this material in the context of the provision of healthcare, where there is potential for therapeutic, diagnostic, scientific, and educational application of the material (to the exclusion of the forensic or artistic application of the material). I will also narrow our focus to the material dimension of separated bodily material (to the exclusion of the informational dimension of the material). The discussion here roams between common law jurisdictions in an attempt to survey landmark cases on the use and storage of bodily material. However, the primary jurisdictional reference point remains the laws of England and Wales, unless otherwise stated.

1

The Ownership of Bodily Material

Mrs Harris gave birth to a baby girl.[1] The baby girl, Rosina, was born with severe birth defects and died after three days. When Mr and Mrs Harris permitted a post-mortem to be conducted on Rosina they required that all organs must be returned so that her body could be buried or cremated whole. During the post-mortem, Rosina's brain, heart, lungs, and spinal cord were removed and retained by Southampton University Hospital. Five years later, Mr and Mrs Harris were informed by letter that the hospital had removed and retained Rosina's organs.

Daniel Carpenter was a sixteen-month-old baby boy.[2] Soon after an operation to remove a tumour near Daniel's brain stem, his condition suddenly deteriorated and he died. Since Daniel's death occurred following an operation, a post-mortem was required. Mr and Mrs Carpenter were initially uneasy about the post-mortem but were assured by the Coroner that the post-mortem would only involve the operation site. During the post-mortem, Daniel's brain was removed and retained by Southampton University Hospital. Samples of the brain tissue were preserved in slides and wax blocks. Fourteen years later, Mr and Mrs Carpenter were informed by letter that the hospital had removed and retained Daniel's brain.

Denise Shorter gave birth to a stillborn baby girl.[3] During Ms Shorter's time in the bereavement ward, the baby, Laura, was kept overnight in the mortuary and brought to Ms Shorter and her husband daily. Mrs Shorter reluctantly agreed to a request for a post-mortem on the condition that everything that was removed from the Laura's body would be returned. Laura's heart and brain were removed and retained by the John Radcliffe Hospital. Eleven years later, Laura's parents were informed by letter that the hospital had removed and retained Laura's heart and brain.

[1] *Re: Organ Retention Group Litigation* [2005] QB 506, 509–10.
[2] *Re: Organ Retention Group Litigation* (n 1) 510–11.
[3] *Re: Organ Retention Group Litigation* (n 1) 511–12.

Controversy, and litigation, followed from these events. The parents of the deceased children were not afforded the ability to possess the (whole) body of their deceased child nor the opportunity to fully determine how and by whom the body of their child is treated and used. The parents asserted a right to *own* the bodies of their deceased children in the sense that they claimed that their ability to possess, and control the use of, the bodies of their deceased child was impinged. To amplify the degree of control that the parents ought to have had, we could say that the bodies of the deceased children were the parents' *property*. The language of property may help us describe the extent to which the parents' rights of possession and control are at the exclusion of any competing interests held by a hospital or any other healthcare institution.

Yet the language of property would seem misplaced in this context. There are important differences between the possession and control of items of property and the possession and control of a newly deceased infant. To appreciate these differences, we need to first conceptually separate *ownership* from *property*. We may, for instance, accept that the parents ought to be able to possess and control the body of their deceased child without having to accept that their legal rights in the bodily material are property rights; we can accept that the body is an owned thing without having to accept that the body is an item of property.

The purpose of this chapter is to clarify the relationship between two distinct concepts: property and ownership. In doing so, we will identify the main inquiries that subsequent chapters will engage in, and identify the basic concepts that subsequent analyses will draw upon. In this chapter I will explain how the concept of 'ownership'—understood as a series of functional entitlements or 'incidents of ownership'—is not a legal concept, and as such, an 'incident of ownership' does not necessitate the application of property law. Since 'incidents of ownership' represent a spectrum of functional relationships between a person and an object or resource, it follows that viewing the law pertaining to bodily material in terms of 'incidents of ownership' has some, but limited, analytical value.

A focus on incidents of ownership is, nonetheless, able to outline the variation in the functional relationships that may arise between a person and an item of bodily material. As I will explain, different 'incidents of ownership' or 'entitlements' represent functionally distinct relationships between a person and an object or resource. It is important to appreciate this functional variation within the concept of ownership since, when we turn to consider why a person ought to be able to exercise the entitlements in bodily material, the explanation or justification will vary depending on the entitlement in question.

The discussion in this chapter provides the backdrop to the two main inquiries that subsequent chapters will undertake. Recall that there are two dimensions of the void in the law left by the 'no property rule'. The common law rule limits the ability for an individual or institution to assert possession or control over an item of bodily material and it prevents any other interests being afforded the legal status of property rights. Since the overall aim of this discussion is to provide a way of filling the void in the law left by the 'no property rule', then the task is twofold: it is necessary to assess which entitlements in an item of bodily material can be justified and which set of legal rules are best suited to protect the exercise of an entitlement in an item of bodily material. Throughout the discussion in this and subsequent chapters, I will therefore be offering both 'theoretical explanations' and 'theoretical justifications'.[4] In terms of theoretical explanations, I aim to explain why property rights 'are as they are'.[5] In particular, why property is distinct from 'ownership' and why property law is distinct from other branches of private law. In terms of theoretical justifications, I aim to 'defend or legitimate'[6] certain claims to the ownership of bodily material (and repudiate others) whilst also evaluating whether property law is the most appropriate branch of law to govern the ownership of bodily material.

I. Incidents of Ownership

There are important similarities between items of property and items of bodily material. Both items are objects that exist physically separate from any particular person. Individuals and institutions can derive value from possessing items of property and items of bodily material, applying such items towards a task, activity or project, and in some circumstances, obtain something valuable in exchange for the item of property or item of bodily material. The intuitive attraction of treating items of bodily material as items of property begins with this observation that items of bodily material appear to function in the same way as items of property.

[4] JL Coleman, *The Practice of Principle: In Defence of a Pragmatist Approach to Legal Theory* (Oxford University Press, 2001) 3: 'There is an important and familiar distinction between theoretical explanations and theoretical justifications. While both can illuminate or deepen our understanding, explanations do so by telling us what the nature of a thing is, or by telling us why things are as they are; by contrast, justifications seek to defend or legitimate certain kinds of things—for example, actions, rules, courses of conduct, practices, institutions, and the like.'

[5] JL Coleman, *The Practice of Principle* (n 4) 3.

[6] JL Coleman, *The Practice of Principle* (n 4) 3.

a. Honoré's Taxonomy

Given this intuition, a popular framework for analysing the legal status of separated bodily material has been Anthony Honoré's taxonomy of ownership.[7] Honoré's seminal essay elucidates the composition of ownership by identifying eleven 'standard incidents' of 'the liberal concept of full individual ownership' which are each constitutive of, but not necessary to, the concept of ownership.[8] These incidents are

the right to possess: to have physical control over the object;

the right to use: to exercise personal use of the object;

the right to manage (or control): to determine how and by whom the object is used;

the right to income (or profit): to derive a benefit from foregoing personal use of the object;

the right to capital: the power to alienate and the liberty to consume the object;

the right to security: insurance that the person will remain owner of the object;

the rights of transmissibility (or transfer): the ability to transfer the rights of ownership to another;

the right to absence of term: presumption of indeterminate length of ownership;

the duty to prevent harm: inability to use the object in harmful ways;

the liability to execution: liable to have the object or asset seized in payment of debt; and

the incident of residuary: rights may expire or be abandoned so as to vest in someone else.

According to Honoré, the concept of ownership consists of a 'bundle' of different relationships between a person and an object. 'Ownership', under this taxonomy, arises when there are a sufficient number of incidents of

[7] AM Honoré, 'Ownership' in *Making Law Bind: Essays Legal and Philosophical* (Clarendon Press, Oxford 1961) 161–93; I Goold, 'Sounds Suspiciously like Property Treatment: Does Human Tissue Fit within the Common Law Concept of Property?' (2005) 7 *UTS L Rev* 62, 63; M Quigley, 'Property and the Body: Applying Honoré' (2007) 33 *J Med Ethics* 631; O Hansson and B Bjorkman, 'Bodily Rights and Property Rights' (2006) 32 *J Med Ethics* 209.

[8] AM Honoré, 'Ownership' (n 7) 162.

ownership.[9] The implication is that there is no necessary or essential feature to the general concept of ownership. Moreover, since different incidents may constitute the concept of ownership, the recognition of one entitlement does not imply the recognition of another. For instance, it does not necessarily follow from the right to transfer bodily material that the transferor can profit from the transfer nor does it follow from the right to profit that the object or asset can be seized in payment of debt.

Given the variety of different contexts in which instances of 'ownership' arise, the malleability of the view that ownership is 'a bundle of rights' is one reason why it has been applied to the use and storage of separated bodily material. As Imogen Goold notes:

> With flexibility comes the ability to adapt to novel forms of property, of which human tissue may be one. This complexity also highlights the conceptual heart of property law, namely that it aims to recognise and regulate conflicting interests and relationships.[10]

Moreover, the bundle view also enables us to deconstruct disputes as to the 'ownership' of bodily material into an organizational structure that identifies the specific aspect or 'incident' of ownership that is being contested. As Muireann Quigley argues:

> [T]he model can help us to move on from abstract disputes about ownership itself and to identify what issues really divide us when considering dilemmas about the use and control of our bodies and their parts that have arisen in the new quasi-commercial climate surrounding human tissue.[11]

It is for these reasons that Honoré's taxonomy 'ownership' can be employed as an organizational structure as it is able to capture a common thread running throughout the different legal disputes whilst still isolating the specific content of the dispute in each case. It is nonetheless important to appreciate what the taxonomy is also unable to capture. As I will explain, it is unable to identify the legal status of an incident of ownership.

[9] AM Honoré, 'Ownership' (n 7), 165; later commentators have adopted Wittgenstein's concept of family resemblance to describe the relationship between the 'incidents of ownership': L Wittgenstein, *Philosophical Investigations* (GEM Anscombe (tr)) (Basil Blackwell, Oxford 1968) 32. See M Quigley, 'Property and the Body: Applying Honoré' (2007) 33 *J Med Ethics* 631.

[10] I Goold, 'Sounds Suspiciously like Property Treatment' (n 7) 63.

[11] M Quigley, 'Applying Honoré' (n 7) 632.

b. Incidents of Ownership and Legal Disputes

There is no doubt that the disputes that have arisen over the use and storage of bodily material have concerned 'incidents of ownership' in bodily material. Let us analyse a series of examples. In terms of *the right to use*, consider the circumstances in *Yearworth v North Bristol NHS* where six men succeeded in their claim in negligence causing damage to property and breach of bailment following the inadequate storage of their extracted semen.[12] They succeeded because the Court of Appeal of England and Wales held that the six men 'had ownership of the sperm for the purposes of their ... claims' on the basis that the men had the Honoréian *right to use* the sperm.[13] Consider also the dispute in *JCM v ANA* between former partners that arose over the use of seven straws of semen that the (then) couple purchased with the intention of undertaking assisted reproductive therapy.[14] As items of property, the Supreme Court of British Columbia divided the seven straws between the two owners on the basis that such a division is consistent with the way in which the couple had agreed to divide their joint assets.[15]

The High Court of Australia in *Doodeward v Spence* recognized an exception to the 'no property rule' in holding that the appellant had the *right to possess* preserved foetal twins for the purposes of their claim in conversion and detinue on the basis that the bodily material had been the subject of 'work or skill'.[16] The Court of Appeal of England and Wales in *R v Kelly* followed the High Court of Australia in holding that the Royal College of Surgeons had the right to possess the preserved body parts for the purposes of s 1(1) of the Theft Act 1968.[17] In comparison to *R v Kelly*, the same court held in *Dobson v North Tyneside Health* that the right to possess the preserved brain of a deceased person did not arise since the then recognized exceptions to the 'no property rule' could not be established on the facts.[18] Namely, the court found that the minimal work or skill was applied to the biological material (the preservation of the brain in fluid) was insufficient for the 'application of skill' exception, and the duty to intern the body had expired upon the burial of the deceased body.[19]

With regards to the right to manage or control bodily material, the parents of the deceased infants (identified at the outset of this chapter) sought

[12] *Yearworth v North Bristol NHS Trust* [2009] EWCA Civ 37.
[13] *Yearworth v North Bristol NHS Trust* (n 12) [28], [45(f)(v)].
[14] *JCM v ANA* [2012] BCSC 584 [95]–[96].
[15] *JCM v ANA* (n 14) [95]–[96].
[16] *Doodeward v Spence* (1908) 6 CLR 406.
[17] *R v Kelly/R v Lindsay* [1998] 3 All ER 741.
[18] *Dobson and another v North Tyneside Health Authority and another* [1996] 4 All ER 474, 479.
[19] *Dobson and another v North Tyneside Health* (n 18) 479.

damages for wrongful interference and negligence causing psychiatric injury in *Re: Organ Retention* following the revelations that the organs of their child had been removed during post-mortem procedures and retained.[20] Under the Human Tissue Act 1961 (HTA 1961), the parents could only determine how and by whom the bodies of their deceased infants were treated through providing generic consent to a post-mortem. No further consent was required for the removal and retention of organs.[21] An action for conversion, or another property action for wrongful interference, was unavailable 'in the absence of such a cause of action in respect of the body of a deceased person being recognized by an English court'.[22] Save for one successful claimant who was able to satisfy the elements of negligence causing psychiatric injury, the claims in *Re: Organ Retention* were unsuccessful in part because the parents' *rights to manage* and control the treatment of their deceased children were constructed narrowly by the HTA 1961.[23]

Subsequent legislation, the Human Tissue Act 2004 (HTA 2004), has expanded the rights of progenitors and family members to control the use of bodily material. Section 1(1) to (3) of the HTA 2004 requires that the storage, use, and removal of human tissue may be lawful provided that it is done with the 'appropriate consent' and pursuant to a Schedule 1 purpose. As a general principle,[24] consent must be given, the consent must correlate with the intended use of the material, and the intended use must be a permissible use as set out in Schedule 1 of the Act. These requirements of lawful possession are fortified by the HTA 2004 offense provisions. Under s 5(1) of the HTA 2004 it is an offense to do any activity listed in s 1(1) to (3) without the 'appropriate consent', and under s 8(1) it is an offense to use or store tissue for a purpose other than a 'qualifying purpose'. Under s 16, the use and storage of human bodies and human tissue also requires a license from the Human Tissue Authority.[25]

[20] *Re: Organ Retention Group Litigation* (n 1).

[21] *Re: Organ Retention Group Litigation* (n 1) [121], Human Tissue Act 1961, s 2(2).

[22] *Re: Organ Retention Group Litigation* (n 1) [161].

[23] *Re: Organ Retention Group Litigation* (n 1) [119]–[127].

[24] There are eight instances under the HTA 2004 where consent is not required for the storage or use of human tissue. The main exception is where the material is stored for a Schedule 1, Part II purpose under the HTA 2004. See J Herring, *Medical Law and Ethics* (4th edn, Oxford University Press, Oxford 2011) 419.

[25] Activities that require licensing under the HTA 2004, s 16(2): '(a) the carrying-out of an anatomical examination; (b) the making of a post-mortem examination; (c) the removal from the body of a deceased person ... of relevant material of which the body consists or which it contains, for use for a scheduled purpose other than transplantation; (d) the storage of an anatomical specimen; (e) the storage ... of—(i) the body of a deceased person, or (ii) relevant material which has come from a human body, for use for a scheduled purpose; (f) the use, for the purpose of public

In lieu of statutory provisions, such as those in the HTA 2004, an important consequence of the 'no property rule' has been to prevent progenitors (and their family members) from controlling the subsequent use of their bodily material. In *Greenberg v Miami Children's Hospital*[26] sufferers of Canavan disease provided tissue and blood samples to a series of researchers. The researchers were then able to isolate the gene that is responsible for Canavan disease. The researchers then obtained intellectual property rights in the isolated gene. The intellectual property right gave the researchers exclusive rights to determine who can access and use the isolated gene, and upon what terms. The exclusivity of the intellectual property right, the donors argued, circumvented the purpose for which they participated in the research by providing bodily material and personal information. However, the donors could not identify any agreement or shared understanding as to the public availability of the isolated gene. The donors' claim in conversion of their human tissue failed since the claimants did not have a proprietary interest in their tissue samples or their genetic information.[27] Similarly, in *Washington University v Catalona*,[28] the donors of tissue had no recognizable property interest in their donated bodily material and therefore retained no ability (beyond the initial provision of consent to donate the tissue) to determine how the material, or the research of the material, is managed.

In terms of the *transferability* of bodily material, the Court of Appeals of California in *Hecht v Superior Court of Los Angeles County*[29] found that a deceased man, who had stored sperm, had 'sufficient decision making authority in relation to the sperm for it to amount to "property" for the purposes of the State's Probate Code'.[30] The semen was then transferred to the deceased's partner as part of his estate. Similarly, in *Bazely v Wesley Monash IVF* a widow sought to obtain the semen of her deceased partner.[31] The deceased had previously extracted and stored the semen for the purposes of assisted reproductive therapy. The Supreme Court of Queensland held that the deceased had the right to possess the semen as property, and through

display, of—(i) the body of a deceased person, or (ii) relevant material which has come from the body of a deceased person.'

[26] *Greenberg v Miami Children's Hospital* 264 F Supp 2d 1064 (SD Fla, 2003), 1066.

[27] *Greenberg v Miami Children's Hospital* (n 26) 1075: '[T]he facts alleged do not sufficiently allege the elements of a *prima facie* case of conversion, as the Plaintiffs have not alleged how the Defendants' use of the Registry in their research was an expressly unauthorized act.'

[28] *Washington University v Catalona* 437 F Supp 2d 985 (Dist Ct Missouri, 2006).

[29] *Hecht v Superior Court of Los Angeles County* (1993) 20 Cal Rptr 2d 275; *Yearworth v North Bristol NHS* (n 12) [40].

[30] *Yearworth v North Bristol NHS* (n 12) [40].

[31] *Bazley v Wesley Monash IVF Pty Ltd* [2010] QSC 118.

the application of the Queensland Succession Act 1981, the right of possession was *transferable* to the widow.[32] In *Re: Edwards* the widow's right to possess the body and bodily material of her deceased husband, including semen extracted after his death, enabled the Supreme Court of New South Wales to narrow the statutory prohibition on 'supplying' semen without the progenitor's consent.[33]

In comparison to these cases where the transfer of bodily material was permissible, the widow in *ex parte Blood*[34] encountered greater limits on the transferability of gametes. The widow sought the right to obtain her deceased husbands' semen which was removed when he was in a comatose state prior to his death. Under the Human Fertilisation and Embryology Act 1990 lawful authority to use the bodily material of the progenitor required the progenitor's consent, and the Court of Appeal of England and Wales held that there was insufficient evidence for the court to infer that the deceased had consented to the use of his reproductive material by his partner.[35]

Finally, consider the two claims that arose in *Moore v Regents of the University of California*.[36] The patient in this case had spleen cells removed as part of his treatment for leukaemia. The physician then developed and patented a cell-line from the patient's spleen cells. The patient successfully claimed that his physician's failure to obtain his informed consent for the use of the extracted cells was in breach of the physician's fiduciary duties to the patient. The patient, by not being informed of the intended use of the bodily material, was deprived of his *right to determine how and by whom* his spleen cells are used and treated.[37] However, the patient was unsuccessful in his claim in conversion for a share of the profits from the licensing of the cell-line.[38] The patient's 'ownership' of his spleen cells does not include the *right to profit* from those cells.

It is therefore possible to view, or map, the law that governs the use and storage of bodily material in terms of 'incidents of ownership'. As I will explain later in this chapter (at pages 26–31), analysing the use and storage of bodily material in terms of 'ownership' helps us appreciate that different 'incidents of ownership' that 'perform different functions and serve different individual and societal interests'.[39] However, as we will now turn to consider,

[32] *Bazely v Wesley Monash IVF Pty Ltd* (n 31) [33].

[33] *Jocelyn Edwards; Re the Estate of the Late Mark Edwards* [2011] NSWSC 478; Assisted Reproductive Technology Act 2007 (NSW), s 21. See also *Re H, AE (No 3)* [2013] SASC 196.

[34] *R v Human Fertilisation and Embryology Authority, ex parte Blood* [1997] 2 All ER 687.

[35] *Ex parte Blood* (n 34) [150].

[36] *Moore v Regents of the University of California* 793 P 2d 478 (Cal 1990).

[37] *Moore v Regents of the University of California* (n 36) 128–35.

[38] *Moore v Regents of the University of California* (n 36) 135–50.

[39] J Christman, 'Distributive Justice and the Complex Structure of Ownership' (1994) 23 *Philos Public Aff* 225, 226.

the application of the ownership taxonomy to bodily material explains next to nothing about the *legal status* of bodily material.

II. Ownership Is Not a Legal Concept

The above legal disputes have been about ownership insofar as they represent disagreement about the right to possess, use, control, transfer, or profit from bodily material. However, the recognition of these incidents of ownership in bodily material tells us very little about the legal status of bodily material. The Court of Appeal of England and Wales in *Yearworth v North Bristol NHS*, whilst applying Honoré's taxonomy of ownership, nonetheless recognized that 'the concept of ownership is no more than a convenient global description of different collections of rights held by persons over physical and other things'.[40] The problem with the 'bundle view' of ownership is that the law does not afford protection of 'incidents of ownership'. As William Swadling explains, 'it is title and not ownership which English law protects. Despite what the layman might think, there is no concept of ownership in English law'.[41]

In order for an incident of ownership to be an incident of ownership that is legally protected, the assertion of ownership needs to be formulated or packaged as a particular legal action. In other words, the functional relationship between the owner and the owned-thing needs to be placed within the context of a legal relationship between the rights-holder (the owner) and the duty-bearer.

a. Incidents of Ownership and Legal Actions

Consider, from the above examples of the 'ownership' of bodily material, how all the claims to ownership are packaged in a particular legal form:

the patients in *Yearworth v North Bristol NHS* succeeded in their claims for breach of bailment and negligence;

as items of property the straws of semen in *JCM v ANA* were divided in accordance with an agreement over the division of relationship property;

the High Court of Australia in *Doodeward v Spence* recognised claims in conversion and detinue;

the Court of Appeal of England and Wales in *R v Kelly* applied the Theft Act;

[40] *Yearworth v North Bristol NHS* (n 12) [28].

[41] W Swadling, 'Unjust Delivery' in A Burrows and A Rodger (eds), *Mapping the Law: Essays in Memory of Peter Birks* (Oxford University Press, Oxford 2006) 281.

the parents' claim in wrongful interference in *Re: Organ Retention* was unavailable due to s 2(2) of the Human Tissue Act 1961 and the 'no property rule';

the donors in *Greenberg v Miami Children's Hospital* were unable to assert the right to control their tissue through actions in conversion because of the 'no property rule';

the ability for the widows in *Bazely v Wesley Monash IVF* and in *Hecht v Superior Court of Los Angeles County* to obtain the semen of their partners was possible through the application of succession law;

the ability for the widow in *ex parte Blood* to obtain the semen of her deceased partner was initially prevented through statutory provisions;

the widow in *Re: Edwards* was able to assert her property rights as a way of avoiding statutory restrictions of the 'supply' of gamates without progenitor consent;

the infringement of the patient's right to manage the use of his spleen cells in *Moore v Regents of the University of California* was a breach of fiduciary duties.

The disputes were all ultimately concerned with ownership, yet in order to assert an incident of ownership in bodily material, the assertion had to be packaged as a particular legal claim or action. Even for the claimants who were able to satisfy an exception to the 'no property rule', they were nonetheless required to satisfy the elements of their legal claim. Moreover, as these examples also illustrate, such legal claims may also confront limits imposed by neighbouring branches of law.

The need for a particular legal action to vindicate the incident of ownership is a need that arises in all instances of ownership. As Honoré himself observed, legal remedies such as ejectment and specific restitution protect an owner's exclusive control so as to uphold the owner's right to possession.[42] Similarly, legal rules and procedures authorize an entitlement-holder to license and consent to acts regarding the object, make contracts regarding the object and transfer the legal title of the object.[43] These legal actions enable the owner to exercise incidents of ownership, such as the ability to determine how and by whom the object is used, to benefit from foregoing personal use and to transfer the object.

The important point here is to appreciate the main limitation of using ownership as an analytical tool. That is, incidents of ownership are incidents of legal rights only insofar as the law imposes duties on others that protect the functional relationship between the owner and the owned-thing. As we will eventually identify (in chapter five), different legal rights protect the relationship between the owner and the owned-thing in structurally different ways. Very little is achieved by placing the law within the ownership

[42] AM Honoré, 'Ownership' (n 7) 166.

[43] AM Honoré, 'Ownership' (n 7) 176.

taxonomy. When we do so, we are merely identifying that a person has an entitlement in an object or resource without knowing anything about how the legal system protects the particular entitlement.

Consider the legal status of embryos as a further example. In *Evans v Amicus Healthcare*,[44] Ms Evans and Mr Johnson extracted gametes and created embryos prior to surgery on Ms Evans' ovaries since the surgery could (and did) deprive her of her natural fertility. The right of the healthcare authority to possess the embryos, the right of the progenitors of the embryos to determine how, and by whom, any embryo is used, was set out in the Human Fertilisation and Embryology Act 1990 (HFEA). The Court of Appeal of England and Wales held that since the HFEA drew clear distinctions between acts of creation, storage, and use, the policy of the Act was to ensure continuing consent of both parties from the commencement of treatment to the point of implant.[45] As a result, Mr Johnson was 'entitled to withdraw his consent with the effect that withdrawal was to prevent both the use and continued storage of the embryo fertilized by the sperm'.[46] The disputes as to the implantation of the fertilized embryo were resolved without consideration of property law.

The Court's prioritization of Mr Johnson's interest in the fertilized embryo led to submissions in the Court of Appeal[47] and the European Court of Human Rights[48] that the requirement of continued bilateral consent for the use and storage of the fertilized embryo was a violation of Ms Evans' right to respect for her private and family life.[49] Both courts recognized that Ms Evans' Article 8 right had to be viewed against Mr Johnson's competing Article 8 right. Since 'motherhood could surely not be forced on Ms Evans and likewise fatherhood surely cannot be forced on Mr Johnson',[50] it was held that Ms Evans' right 'to respect for the decision to become a parent in the genetic sense should [not] be accorded greater weight than [Mr Johnson's]

[44] *Evans v Amicus Healthcare Ltd and others* [2004] EWCA (Civ) 727.

[45] *Evans v Amicus Healthcare* (n 44) [36].

[46] *Evans v Amicus Healthcare* (n 44) [41].

[47] *Evans v Amicus Healthcare* (n 44) [111].

[48] *Evans v United Kingdom* (2008) 46 EHRR 34 (Grand Chamber).

[49] European Convention on Human Rights, Article 8: '(1) Everyone has the right to respect for his private and family life, his home and his correspondence. (2) There shall be no interference by a public authority with the exercise of this right except such as is in accordance with the law and is necessary in a democratic society in the interests of national security, public safety or the economic well-being of the country, for the prevention of disorder or crime, for the protection of health or morals, or for the protection of the rights and freedoms of others.'

[50] *Evans v Amicus Healthcare* (n 44) [111].

right to respect for his decision not to have a genetically related child with her'.[51]

Compare the reliance on statutory rights in *Evans v Amicus Healthcare* with the recognition of property rights in *CC v AW*.[52] In the latter case AW provided sperm to assist CC in the conception of children. This provision of sperm was held to be an 'unqualified gift' that was 'extended as a courtesy to a friend'.[53] Fertilized embryos were formed (from AW and CC's gametes) and CC conceived twins. AW refused to consent to the release of the remaining four embryos to CC. The Alberta Queen's Bench held that the 'remaining fertilised embryos remain [CC's] property. They are chattels that can be used as she sees fit'.[54]

The decision of the Supreme Court of Tennessee in *Davis v Davis* provides a third point of comparison.[55] The Supreme Court sought to avoid the application of property law whilst still attempting to recognize entitlements in the embryo. In *Davis v Davis*, embryos remained after unsuccessful attempts at *in vitro* fertilization. Following the dissolution of the relationship, Ms Davis sought to have the embryos donated to another couple whilst Mr Davis sought to have the embryos destroyed. In considering whether the embryos were items of property, the Court held that

> any interest that [Ms] Davis and [Mr] Davis have in the pre-embryos in this case is not a true property interest. However, they do have an interest in the nature of ownership, to the extent that they have decision-making authority concerning the disposition of the pre-embryos.[56]

To resolve the dispute, the Supreme Court turned to contract law and then the constitutional right to privacy. The Court held, 'as a starting point' any agreement regarding the 'disposition of any untransferred preembryos in the event of contingencies... should be presumed valid and should be enforced as between the progenitors'.[57] Absent any agreement or prior indication of preferences, the Court held that 'the right of procreation is a vital part of an individual's right to privacy'.[58] In the same balancing exercise that the European Court of Human Rights was asked to engage in, the Supreme Court of Tennessee held that 'the party wishing to avoid procreation should prevail'.[59] Yet, unlike the European Court in *Evans v UK*, the Supreme Court

[51] *Evans v United Kingdom* (n 48) para 90. See M Ford, '*Evans v United Kingdom*: What Implications for the Jurisprudence of Pregnancy?' (2008) 8 *HRL Rev* 171, 173.

[52] *CC v AW* 2005 ABQB 290.

[53] *CC v AW* (n 52) [20].

[54] *CC v AW* (n 52) [21].

[55] *Davis v Davis* 842 SW 2d 588 (1992).

[56] *Davis v Davis* (n 55) 597.

[57] *Davis v Davis* (n 55) 597.

[58] *Davis v Davis* (n 55) 600.

[59] *Davis v Davis* (n 55) 604.

added the proviso that non-procreation prevails 'assuming that the other party has a reasonable possibility of achieving parenthood by means other than use of the preembryos in question'.[60]

In all three cases, the embryos were owned things (insofar as they were the subject of incidents of ownership). The three cases differ, in part, because of the different branches of law that were applied to address the legal dispute. In *Evans v Amicus Healthcare*, Mr Johnson was 'entitled' under statue 'to withdraw his consent with the effect that withdrawal was to prevent both the use and continued storage of the embryo fertilized by the sperm'.[61] The entitlements in the embryo in *Evans,* and the relative priority of these entitlements, were governed by a structure of statutory rights and duties. In *CC v AW,* CC's right to use the embryo was governed by property law and Mr AW 'was not in a position to control or direct their use in any fashion'.[62] In *Davies v Davies*, the right to use the embryos, and the relative priority of such rights, were determined with reference to contractual agreement and then a constitutional right to privacy.[63]

Since there are different ways in which the law can govern the use of embryos, the exercise of the right to possess, control, or use embryos does not equate to the exercise of property rights. There remain further questions as to whether property law is the branch of law that is best suited for the task of protecting the possession and use of embryos, or any other item of bodily material.

b. Twin Notions of Property

Property law represents a particular structure of rights and duties that may be, and most often is, employed to provide legal protection for the possession and use of objects and resources. Since property is by far the most common means of protecting entitlements in objects and material resources we often use the terms 'ownership' and 'property' interchangeably. They are not, however, interchangeable concepts and it is important that we clarify the relationship between ownership and property. The distinction can be understood as follows: whereas ownership represents the functional relationship between a person and a thing, property law represents a particular structure of rights and duties that protects the ownership relationship in a particular way.

Hence, all instances of property rights include an entitlement (or incident of ownership) in an object or resource. An incident of ownership is a necessary condition of a property right. However, as James W Harris explains, 'the

[60] *Davis v Davis* (n 55) 604.
[61] *Evans v Amicus Healthcare* (n 44) [41].
[62] *CC v AW* (n 52) [21].
[63] *Davis v Davis* (n 55) 604.

essentials of a property institution are the twin notions of trespassory rules and the ownership spectrum'.[64] To understand property law is to understand both the way in which property protects entitlements in objects or resources ('trespassory rules') and the various entitlements that a person may have in an object or resource ('the ownership spectrum'). It follows that identifying an incident of ownership in bodily material does not amount to establishing a property right in the bodily material. An incident of ownership is *not a sufficient* condition of a property right.

Rights of possession, management, use, and transfer can be exercised in bodily material without the use of property law. As we have already seen, the requirement of consent and fiduciary duties imposed on physicians or medical professionals can facilitate a patient's authorization for whether, and for what purposes, their bodily material is used. Statutory rights and duties are able to govern how, and by whom, bodily material is stored, is used, and transferred. Duties in negligence and confidentiality also have the potential to protect the 'ownership' of bodily material by imposing duties on those who interact with the 'owner' of the item of bodily material itself. The key point is that the exercise of entitlements or 'incidents of ownership' does not necessarily amount to the exercise of property rights.

Yet, it has so far been a common mistake to conflate incidents of ownership with property. For instance, Radhika Rao argues that:

> If society permits a part of the body to be separated from the person and alienated to others or seized by the state, that part of the body constitutes 'property', regardless of the legal label attached to it.[65]

However, statutory provisions, common law rules beyond property law, and equitable duties are able to facilitate the transfer of bodily material to another, or permit public authorities to obtain, or seize, bodily material. By permitting this, it does not follow that there is a recognition of property in body parts, the law merely recognizes entitlements in body parts. It is a further question, and not just a question of labelling, how these entitlements are legally protected.

Goold adopts Honoré's eleven incidents of ownership and shows how it is possible to fit separated bodily material into Honoré's framework, to conclude that 'human tissue is aptly suited to having property status, and that... [m]ore generally... at the conceptual level, there are few legal difficulties with viewing human tissue as property'.[66] Quigley also adopts the

[64] JW Harris, *Property and Justice* (Clarendon Press, Oxford 1996) 5.

[65] R Rao, 'Property, Privacy and the Human Body' (2000) 80 *BU L Rev* 359, 460.

[66] I Goold, 'Sounds Suspiciously like Property Treatment' (n 7) 62.

taxonomy as a theory of property and asserts that, 'with the support of the most complete theory of property that we can have', we can say when it comes to my body that 'I own that or that is my property'.[67] However, by identifying incidents of ownership in separated bodily material, we are merely identifying a functional relationship between the person and the object; that the 'owner' is able to possess, use, control, or transfer the object. It does not tell us anything about how the law protects the relationship between a person and the object. To answer the legal questions, about how the law protects the relationship between a person and object, we need to identify the particular structural features of the different branches of law that can protect 'entitlements in things'.

Therefore, to say that someone is an owner of a thing, is merely to say that the individual has particular entitlements in the thing that the law—*somehow*—protects. As subsequent chapters will explain, each legal action, and each branch of private law, has a distinct legal structure that may be well suited, or ill-equipped, to protect the 'ownership' of bodily material. This raises a series of important questions, such as: what is distinct about property law as a means of protecting incidents of ownership in objects and resources? What other branches are capable of protecting incidents of ownership? What are the relative strengths and weaknesses of these different legal categories when considering the use and storage of bodily material? We return to these questions in chapters four, five, and six.

However, before we engage in these questions as to the suitable means of legal protection of entitlements in bodily material, we need to first identify both the entitlements that ought to arise in bodily material and the justification for why such entitlements ought to arrive. This is because we need to obtain a sense of the relationships between a person and an item of bodily material that ought to obtain legal protection before we discuss the appropriate structure of the legal protection. As we shall now turn to consider, the ownership taxonomy is able to assist us in this primary task of identifying who should be able to exercise entitlements in bodily material, and why.

III. Ownership Is a Series of Functional Relationships

Although the 'bundle view' of ownership is unable to clarify the *legal* status of bodily material, viewing disputes over the use and storage of bodily material in terms of the different incidents of ownership does illustrate how

[67] M Quigley, 'Applying Honoré' (n 7) 633.

different types of ownership relationship fall within the global idea of 'ownership'. The possession of the preserved body parts, the use of gametes in assisted reproduction therapy, the control over the body of a deceased child, the transfer of possession of semen from a healthcare institution to a widow, and the profit from the use of a spleen cell, all represent different types of relationships between a person and an item of bodily material. Although they are all 'incidents of ownership', there exists between the incidents of ownership important functional differences. Each incident of ownership or entitlement enables different preferences and choices, serve different interests, and raise different concerns.

a. Functional Variation

Before we consider whether an individual or institution ought to own, or have an entitlement in, bodily material, we need to first isolate the functional relationship that each incident of ownership or entitlement represents.

The entitlement to *possess* an object (in functional, not legal, terms) only enables the right-holder to determine the physical location of the object. The entitlement to possess therefore concerns a simple relationship between the entitlement-holder and the location of the object. Healthcare institutions (such as North Bristol NHS, Amicus Healthcare, Wesley Monash IVF, etc.) may obtain the entitlement to possess bodily material. In such instances, the institutions may only obtain a bare entitlement to possess that is not extended to include other entitlements.

The entitlement to *use* an object enables a relationship between 'an owner's set of preferences ... and the general state of the object owned'.[68] It enables the entitlement-holder to change the features or characteristics of the object or to apply the object or resource to a particular task. The claimants in *Yearworth v North Bristol NHS*, for instance, had the right to determine the general state of their semen in a way that aligns with their preferences and choices. This right to use the semen as part of assisted reproductive procedure was infringed by the inadequate storage of the semen. A dimension of the entitlement to determine the state of the object is that an entitlement-holder may wish to destroy the object. Hence, in *Evans v Amicus Healthcare, CC v AW,* and *JCM v ANA*, disputes arose between the positive exercise of a use entitlement and the claim to negate or prevent any use of the embryos or gametes. Healthcare institutions also obtain the right to use bodily material provided that they satisfy the statutory preconditions of being licensed to use bodily

[68] J Christman, 'Distributive Justice and the Complex Structure of Ownership' (n 39) 232.

material, obtaining appropriate consent and using the bodily material pursuant to a qualifying purpose.[69]

As we move further through the taxonomy, a more complicated series of relationships begins to emerge. The entitlement to *manage* or *control* an object adds the power to determine the way that others possess or use an object. Although a person may be able to determine the physical location of the object and (to some extent) the general state of the object, these entitlements may be subject to the superior entitlement of another person is able to determine how and by whom the object is used. In terms of the retention of bodily material by healthcare authorities for purposes beyond the treatment of the progenitor, often the right to control is narrowed to a right to authorize particular uses of the bodily material. For instance, in response to the controversy that followed the events in *Re: Organ Retention* and other similar practices,[70] the Human Tissue Act 2004 strengthened the consent requirements for the removal, storage, and use of human tissue. As a general principle, consent must be given, the consent must correlate with the intended use of the material, and the intended use must be a permissible use as set out in Schedule 1 of the Act.[71] A progenitor (or family member) may therefore determine, in general terms, how their tissue is used by a healthcare authority.

The entitlement *to transfer* enables the entitlement-holder to forgo their set entitlements in favour of another person. The entitlement-holder, in exercising the right to transfer, has the ability to change the ownership relationship itself, from the transferor (and the object) to the transferee (and the object). This entitlement is an allocative power insofar as the transferor can determine how an object or resource is allocated between potential entitlement-holders. Courts have recently been willing to recognize the ability to transfer bodily material to be part of the bundle of posthumous entitlements that a person may have over their body or bodily material. For instance, the application of the succession legislation in *Bazely v Wesley Monash IVF* enabled the transfer of entitlements in stored semen from the deceased to his widow. This is an indirect way of ultimately recognizing a much simpler functional relationship. Since the progenitor and the widow were co-participants in an intended therapy, and the bodily material was extracted for this therapy, it is arguable

[69] Human Tissue Act 2004, s 1(1)–(3), s 8(1), s 16; Human Fertilisation and Embryology Act 1990, s 41, ss 1–4 of Schedule 3, ss 5–8 of Schedule 3.

[70] See K Liddell and A Hall, 'Beyond Bristol and Alder Hey: The Future of Regulation of Human Tissue' (2005) 13 *Med L Rev* 170.

[71] HTA 2004, ss 1–5; See J Herring, *Medical Law and Ethics* (n 24) 419.

that upon the death of the progenitor the entitlements of the progenitor ought to vest in the widow. This is an 'incident of *residuary*'.[72]

The entitlement *to profit* enables the entitlement-holder to transfer their entitlements in exchange for monetary gain. Not only is the entitlement-holder able to change the ownership relationship, the entitlement-holder can financially benefit from forgoing entitlements in the object or resource. This is a complicated relationship that involves four components: the transferor, the object or resource, the recipient, and the financial benefit. It is important to note that, for the entitlement to profit, the content of the financial benefit is determined by the transferor and recipient against a wider distributional background (discussed further in chapter three, at pages 81–2).

In a number of instances, the entitlement to profit may be reduced to an entitlement to be compensated. In such instances, the measure of the financial benefit is determined by a legal or regulatory standard rather than by the parties themselves. In terms of such financial benefits, s 32 of the Human Tissue Act 2004 prohibits the commercial transfer of 'controlled material'[73] and s 12(1) of the Human Fertilisation and Embryology Act 1990 prohibits the payment of any money 'given or received in respect of any supply of gametes or embryos'.[74] Yet, the Human Fertilisation and Embryology Guidelines permit payments reimbursing 'all reasonable expenses incurred ... in connection with donating gametes or embryos' and payments 'compensating the donor for their expenses and loss of earnings'.[75] Moreover, s 32(7) of the Human Tissue Act 2004 provides that it is not an offence under the Act to make payments to a donor of human tissue to compensate for expenses incurred or loss of earnings.[76] A progenitor may therefore exercise the ability

[72] AM Honoré, 'Ownership' (n 7) 179.

[73] HTA 2004, s 32.

[74] Human Fertilisation and Embryology Act 1990, s 12(1): 'The following shall be conditions of any license granted under this Act ... (e) that no money or other benefit shall be given or received in respect of any supply of gametes, embryos or human admixed embryos unless authorised by Directions ...'

[75] Human Fertilisation and Embryology Authority, *Code of Practice* (2009) [13.2]–[13.4]: 'Gamete and embryo donation Directions: Centres may compensate sperm donors a fixed sum of up to £35 per clinic visit. Centres may compensate egg donors a fixed sum of up to £750 per cycle of donation. Where a prospective egg donor does not complete the cycle, the centre may compensate the egg donor on a "per clinic visit" basis.'

[76] HTA 2004, s 32(7): 'References in subsections (1) and (2) to reward, in relation to the supply of any controlled material, do not include payment in money or money's worth for defraying or reimbursing—(a) any expenses incurred in, or in connection with, transporting, removing, preparing, preserving or storing the material, (b) any liability incurred in respect of—(i) expenses incurred by a third party in, or in connection with, any of the activities mentioned in paragraph (a), or (ii) a payment in relation to which subsection (6) has effect, or (c) any expenses or loss of earnings incurred by the person from whose body the material comes so far as reasonably and directly attributable to his supplying the material from his body.'

to transfer bodily material, and may also be financially compensated for the transfer of bodily material, but may not exercise a fully-fledged right to profit from their bodily material.

Each entitlement therefore represents a distinct functional relationship between the person and object, ranging from the person being able to: determine the physical location of the object, determine the general state of the object, determine how and by whom the object is treated, forgo the entitlements in the object in favour of another person, or financially benefit from (or be compensated for) the transfer of entitlements to another person. These are functionally distinct relationships; the relationship between the rights-holder and the object of resource differs depending on the permitted activities that can be undertaken and the essential components of the ownership relationship.

b. The Ownership Spectrum

This variation in ownership has been described by Harris as a spectrum that ranges between 'mere property' and 'full blooded ownership'.[77] Mere property amounts to 'some open ended set of use-privileges and some open ended set of powers of control over uses made by others'.[78] Full-blooded ownership entails 'prima facie, unlimited privileges of use … unlimited powers of control and transmission'.[79] Harris suggests that 'property', in its fullest sense, has the twin functions of 'governing both the use of things and the allocation of social wealth'.[80] Hence, not only is the ('full blooded') owner able to determine the state of the object, the owner also controls, and can realize, the value of the object.

In a similar vein, John Christman suggests that since 'different aspects of property rights tend to perform different functions and serve different individual and societal interests',[81] entitlements can be divided into categories. Christman draws an important distinction between 'control rights' and 'income rights'. Control rights enable the rights-holder to be the 'primary arbitrator over what is to be done with a thing'[82] whereas income rights enable the rights-holder to derive a benefit in exchange for her relinquishing one or more of her control rights.[83]

[77] JW Harris, *Property and Justice*, (n 64) 27.
[78] JW Harris, *Property and Justice*, (n 64) 27.
[79] JW Harris, *Property and Justice*, (n 64) 30.
[80] JW Harris, *Property and Justice* (n 64) 26.
[81] J Christman, 'Distributive Justice and the Complex Structure of Ownership' (n 39) 226.
[82] J Christman, 'Self-Ownership, Equality and the Structure of Property Rights' (1991) 19 *Polit Theory* 28, 29.
[83] J Christman, 'Self-Ownership, Equality and the Structure of Property Rights' (n 82) 29.

The reason for identifying the functional differences between the 'incidents of ownership' is to appreciate how the task of justifying 'incidents of ownership' varies. It is because different incidents of ownership perform different functions, different incidents of ownership will serve different interests and raise different concerns. When we assess whether a person ought to be able exercise an incident of ownership in an item of bodily material, the assessment will vary depending on the particular incident of ownership in question.

For instance, although we may be able to identify sufficient reasons why a person ought to be able to possess (determine the physical location) or even use (determine the general physical state) an object or resource, we ought not to assume that such reasons are sufficient to justify or explain why the same person ought to be able to forgo these entitlements in favour of another person or financially benefit from the transfer. Hence, even if we can explain why a person ought to be able to possess and control an object or resource, a further explanation of justification is required to explain why the person ought to have further powers and privileges with regards to the object or resource.

IV. Justifying Ownership

Recall that there are two dimensions of the void in the law left by the 'no property rule'. The first dimension is that it limits the ability of an individual or institution to assert possession or control over an item of bodily material. The second aspect is that it prevents any other interests being afforded the legal protection of property rights. The first task of providing a way of filling this void is to identify the entitlements that a person (an institution or individual) ought to be able to exercise in bodily material. There are, however, a variety of bases upon which we may justify or explain why a person ought to able to exercise an entitlement in an object or resource. Here, I will briefly draw a distinction between two different justificatory bases for the recognition of entitlements in bodily material, at which point we will have isolated the basic building blocks for undertaking the first task of identifying the entitlements that a person ought to be able to exercise in bodily material.

a. Range of Justifications

One intuitive explanation for why a person ought to be able to possess, use, control, transfer, or even profit from bodily material is that the bodily material was originally *theirs* in the first place. Incidents of ownership in bodily material, according to this intuition, may represent a 'natural extension' of the right to bodily integrity.[84] As we have seen in *Williams v Williams, Moore v Regents of the University of California, Greenberg v Miami Children's Hospital*, and *Washington University v Catalona*, the 'no property rule' has prevented any legal recognition of an extended right to bodily integrity.

Instead, the exceptions to the 'no property rule' permit entitlements in bodily material in circumstances where the entitlement-holder purposely interacts or engages with the bodily material. For instance, the High Court of Australia in *Doodeward v Spence* held that where a person has 'by the lawful exercise of work or skill so dealt with a corpse in his lawful possession that it has acquired some attributes differentiating it from a mere corpse' that person acquires the 'right to possession of the corpse or part'.[85] The Court of Appeal of England and Wales used a similar justification in *R v Kelly* finding that 'where a corpse, or part of a corpse, had undergone a process of application of skill designed to preserve it for medical or scientific examination, it ... became property for the purposes of the Theft Act 1968'.[86] Although not requiring 'some attribute to differentiate it from a mere corpse' the same court had previously held in *Dobson v North Tyneside Health Authority* the preservation of a brain in fluid to be insufficient 'application of skill' to recognize a right to possess the brain.[87] This notion of the recognition of ownership entitlements through the application of human skill also appears in s 32(9)(c) of the Human Tissue Act 2004. Section 32(9)(c) states that 'controlled material' that 'is the subject of property because of the application of human skill' is exempted from the prohibition of commercial dealings in human tissue.[88]

With regards to the work or skill exception to the 'no property rule', the Court of Appeal of England and Wales in *Yearworth v North Bristol NHS* was not content to see the law develop in this way.[89] Rather, the court justified the patients' right to use the semen on the basis that 'by their bodies, they alone generated and ejaculated the sperm' and the 'sole object was that

[84] R Hardcastle, *Law and the Human Body: Property Rights, Ownership and Control* (Hart Publishing, Oxford 2009) 147.

[85] *Doodeward v Spence* (n 16) 414.

[86] *R v Kelly* (n 17) 750.

[87] *Dobson and another v North Tyneside Health* (n 18) 479.

[88] HTA 2004, s 32(1). The fact that the HTA 2004 confines the work or skill exception to 'controlled material' gives rise to an important interpretative point. As Hardcastle explains, the original clause (clause 29 of the Human Tissue Bill) prohibited commercial dealing in all 'relevant bodily material'. The definition of 'relevant bodily material' included any bodily material which consisted of, or included, human cells. The prohibition against commercial dealings in the legislation was limited to 'controlled material', which is defined in s 32(8) as consisting of, or including, human cells, removed from the human body, that are *intended to be used for the purpose of transplantation*.

[89] *Yearworth v North Bristol NHS* (n 12) [45(d)].

it might be used for their benefit'.[90] There was, for the Court of Appeal, an important connection between the progenitors and their semen that was not captured by the 'work or skill' rule. The underlying basis for the exception to the 'no property rule' identified in *Yearworth v North Bristol NHS* is that a connection between the progenitors and their separated bodily material is retained, or a connection arises, through the progenitors intentionally causing the separated state of the bodily material.

Recall how the contested use of the embryo in *Evans v Amicus Healthcare* engaged the human rights of the progenitors.[91] In terms of the use of gametes we can appreciate how the progenitor's control of their reproductive material engages the right of respect for their private and family life. More broadly, to the extent that human rights norms protect moral properties that each person has by virtue of their humanity,[92] we can also appreciate that the use and control of bodily material may also engage further moral properties that each person has by virtue of their humanity. The engagement with such moral properties may provide a justificatory basis for the rights of use and control.

Note how these justifications focus on the connection between some intentional action, feature or attribute of the entitlement-holder and the bodily material. It is because the bodily material forms part of the entitlement-holders bodily integrity, was generated and extracted for the entitlement-holder's benefit, or engages moral properties that the entitlement-holder has by virtue of their humanity, that entitlements arise in the bodily material.

Other explanations or justifications for the recognition of entitlements in bodily material focus on a different set of considerations. For instance, the Nuffield Council for Bioethics suggests that where the current levels of donation of bodily material creates a 'health need', the appropriate response is to incentivize the donation or transfer of bodily material by permitting different levels or 'rungs' of compensation, reimbursement, remuneration, reward, or purchase. The appropriate level of 'intervention' is determined by the need for the resources (with reference to other public policy concerns).[93] It may be

[90] *Yearworth v North Bristol NHS* (n 12) [45(f)].

[91] *Evans v Amicus Healthcare* (n 44) [111]; *Evans v United Kingdom* (n 48).

[92] J Tasioulas, 'On the Nature of Human Rights' in G Ernst and J-C Heilinger (eds), *The Philosophy of Human Rights: Contemporary Controversies* (Berlin; Boston: De Gruyter, 2012) 17, 26; T Endicott, 'The Infant in the Snow' in T Endicott, J Getzler, and E Peel (eds), *Properties of Law* (Oxford University Press, Oxford 2006) 348, pp 348, 362; J Griffin, *On Human Rights* (Oxford University Press, Oxford 2008) pp 8–9.

[93] Nuffield Council on Bioethics, *Human Bodies: Donation for Medicine and Research* (London: Nuffield Council on Bioethics, 2011) 167–70. The suggestion by the Nuffield Council is that where the 'health need' is not being met by current forms of intervention, a higher level of intervention may be justified, provided there is consideration as to: the welfare of the donor, the welfare of other closely concerned individuals, the potential threat to existing donation systems,

possible, therefore, to justify the exercise of an entitlement in bodily material on the basis that such entitlements are allocated in accordance to distributional patterns desired by society or social institutions. There may be a societal interest in encouraging medical research or increasing the availability of bodily material sufficiently to justify a right to transfer bodily material and the right to be financially rewarded from the transfer of bodily material.

Consider further the storage of human tissue—without the patient's consent—that is no longer needed for the patient's own care or treatment by a healthcare authority for the purposes of clinical audit, education, or training relating to human health, or monitoring of health services.[94] The ability of healthcare institutions to be able to retain and store such bodily material (without the progenitor's consent) may be justified with reference to the benefit that all patients obtain from having an audited, monitored healthcare system that has educated and trained healthcare professionals.

In the same vein, proponents of the routine recovery of organs from the dead argue that the lives that can be prolonged and saved through the transplantation of bodily material[95] can justify the right of healthcare institutions to possess and use the bodily material of a deceased person (uncircumscribed by the wishes of the deceased or the deceased's family members).[96] These justifications for increased 'public intervention' to encourage the provision of bodily material, the retention of bodily material without the progenitor's consent, and the routine recovery of organs share the same structure. The entitlement in bodily material (compensation, retention, routine recovery) is justified by connecting the entitlement with the consequences that follow from the exercise of the entitlement in bodily material.

b. Pre-Social and Social Analyses

Although there are a variety of different possible justifications for the recognition of entitlements in bodily material, the normative force of a justification either follows from respect of a morally relevant intentional act, attribute, or feature of a person, or the normative force stems from the value of an outcome or set of consequences that follow from the exercise of an entitlement

the risk of increasing social inequalities, and the professional responsibilities of the health professionals involved.

[94] HTA 2004, Schedule 1, Part II Purpose; J Herring, *Medical Law and Ethics* (4th edn, Oxford University Press, Oxford 2011) 419.

[95] For instance, A Spital and JS Taylor, 'Routine Recovery of Cadaveric Organs for Transplantation: Consistent, Fair, and Life-Saving' (2007) 2 *Clin J Am Soc Nephrol* 300, 303.

[96] *cf* HTA 2004, s 4.

in bodily material. Hence, we can—as Joshua Getzler has—distinguish between two types of analysis:

> There is a notion of property as presocial, a natural right expressing the rights of persons which are prior to the state and law, this being the view of Hugo Grotius, Samuel von Pufendorf, John Locke, Immanuel Kant, and Georg W. F. Hegel; and there is a notion of property as social, a positive right created instrumentally by community, state, or law to secure other goals—the theory of Thomas Hobbes, David Hume, Adam Smith, Jeremy Bentham, Emile Durkheim, and Max Weber.[97]

To appreciate this distinction, recall the use of the spleen cells in *Moore v Regents of the University of California*. The majority of the Supreme Court of California considered that there were two policy considerations 'of overriding importance'.[98] The first was the 'protection of a competent patient's right to make autonomous medical decisions'. The patient's right to control the use of his bodily material also concerned values of 'dignity' and 'privacy'.[99] For the majority, this consideration was adequately addressed though the imposition of equitable duties on the physician. The second consideration was the concern that the extension of tort liability (through the finding that the physician converted the spleen cell) 'threatens to destroy the economic incentive to conduct important medical research'.[100] It was this second concern that was a major factor in the majority's decision to deny the claim in conversion. The Supreme Court engaged in a *presocial analysis* to explain why the patient had a right to be informed about the intended use of his spleen cells and engaged in a *social analysis* when considering whether to extend tort liability to bodily material.

Akin to Getzler's distinction, other commentators have articulated similar ways of isolating the concept of a pre-social right. For HLA Hart, a natural right is a right 'which all men have if they are capable of choice; they have it *qua* men and not only if they are members of some society or standing in some special relation to each other'.[101] John Rawls gives a slightly wider account of natural rights as claims which 'depend solely on certain natural attributes the presence of which can be ascertained by natural reason pursuing common sense methods of inquiry'.[102]

[97] J Getzler, 'Theories of Property and Economic Development' (1996) 26 *J Interdiscip Hist* 639, 641.

[98] *Moore v Regents of the University of California* (n 36) 143.

[99] *Moore v Regents of the University of California* (n 36) 140.

[100] *Moore v Regents of the University of California* (n 36) 141.

[101] HLA Hart, 'Are There Any Natural Rights?' (1955) 64 *Phil Rev* 175–91, 175; J Christman, 'Can Ownership be Justified by Natural Rights?' (1986) 15 *Philos Public Aff* 156, 157–8.

[102] J Rawls, *A Theory of Justice* (Harvard University Press, 1971), 505; J Christman, 'Can Ownership Be Justified by Natural Rights' (n 101) 157–8.

In an attempt to isolate potential pre-social rights to bodily material, in subsequent chapters we will focus on the *subject*. The term is intended to encompass more than the traditional concept of a person's subjectivity pertaining to their 'consciousness, self-consciousness, or reflection'[103] or 'that to which representations are attributed or predicated'.[104] Here the term is used in a broader sense; a person's *subjectivity* is the feature that a person has that distinguishes them from other *objects*. More precisely, a person's subjectivity represents the attributes, qualities or properties that a person has that grounds the moral duties that are owed to that person. I leave the attributes, qualities or properties of the subject as open-ended. Such qualities could be capacity for reason and autonomous action,[105] responsiveness to pain and pleasure, an interest in well-being or preference satisfaction,[106] the capacity to form, to pursue, and potentially to revise a plan of life,[107] or human dignity and an aggregate of objective human thriving.[108] What these attributes, qualities, or properties are, which provide the basis of the moral duties that we owe to others, is not the focus. What we will be interested in is how these components of our subjectivity (however constituted) connect and interact with the world of objects, in particular how our subjectivity relates to, or is connected with, our bodies.

The contention here is only that an entitlement may be justified in a *pre-social analysis* that considers whether an entitlement in a thing can be justified with reference to the attributes, qualities, or properties of the person. It is in this way that duties of respect for the person may extend to respect for their possession, use, or control of their bodily material. Whether entitlements in bodily material can be justified in a pre-social analysis will depend upon whether items of bodily material retain a connection with a person's subjectivity. Alternatively, an entitlement may be justified in a *social analysis*, which considers whether an entitlement in a thing can be justified with reference to the state of affairs that the exercise of the right will produce.

[103] S Critchley, *Ethics-Politics-Subjectivity* (Verso, London 1999) 55.

[104] S Critchley, *Ethics-Politics-Subjectivity* (n 103) 51.

[105] See, for instance: TM Powers, 'The Integrity of Body: Kantian Moral Constraints on the Physical Self' in MJ Cherry (ed) *Persons and Their Bodies: Rights, Responsibilities, Relationships* (Kluwer Academic Publishers, 1999) 213.

[106] See, for instance: W Donner, 'A Millian Perspective on the Relationship between Persons and Their Bodies' in MJ Cherry (ed), *Persons and Their Bodies: Rights, Responsibilities, Relationships* (Kluwer Academic Publishers, 1999) 61.

[107] See, for instance: C Fabre, *Whose Body Is It Anyway? Justice and the Integrity of the Person* (Oxford University Press, Oxford 2006) 28.

[108] See, for instance: C Foster, *Human Dignity in Bioethics and Law* (Hart Publishing, Oxford 2011) 15.

These instrumental justifications use the entitlement *as a means to* produce a desirable state of affairs. As we have seen, such valuable outcomes may include economic incentives for medical research, the wider allocation of bodily material, or the saving or prolonging of the lives of others.

Both pre-social and social analyses represent legitimate ways to justify entitlements in bodily material. But since they draw upon different sets of considerations, it is important to appreciate that an entitlement in bodily material can be justified on different, sometimes competing, bases. Moreover, given the functional variation in different entitlements, some justificatory bases may be unable to justify the full range of entitlements. This raises a series of questions that the next two chapters aim to address, such as: whether there is a connection between a person and their bodily material that is able to justify pre-social entitlements in bodily material? If so, which entitlements can be justified on a pre-social basis? Given the demands on bodily material that follow from a social analysis, how do we calibrate the relative normative weight of conflicting social and pre-social bases of ownership, and are there other limits to the outcome orientated calculus of a social analysis?

V. Chapter Summary

This chapter has isolated the basic building blocks for the analysis that follows. First and foremost, I have sought to clarify the relationship between 'property' and 'ownership'. 'Ownership', I have argued, is a global description of a series of functional relationships between a person and an object or resource. The recognition and exercise of an 'incident of ownership' in bodily material are not, therefore, in any way indicative of the legal status of the bodily material. Viewing disputes as to the use and storage of bodily material in terms of 'incidents of ownership' does, however, assist us in a related inquiry. That is, it enables us to differentiate between the types of relationships that may arise between an object or resources and a person. This variation within the global concept of 'ownership' is relevant to how we assess whether a person ought to be able to 'own', or exercise an entitlement in, bodily material.

It is a further and distinct question as to *how the law protects* the relationship between a person and an object or resource. Property represents a particular way in which the law may protect an incident of ownership or entitlement. To assess whether property rights ought to arise in bodily material requires an assessment of whether property law represents the most appropriate configuration of rights and duties for the governance of the use and storage of bodily material.

There are, therefore, two main tasks for the remaining chapters. We shall turn next to the task of assessing which entitlements in bodily material an individual or institution ought to have in bodily material. This requires an assessment of whether there is a connection between a person and their bodily material that can give rise to a pre-social entitlement, what entitlements can be justified in a social analysis, and how to address the tensions that arise between competing social and pre-social justifications of entitlements in bodily material. The second half of the book is then concerned with the configuration of legal rights and duties that ought to apply to the exercise of entitlements in bodily material. These latter chapters will identify how property law protects entitlements in objects and resources, how property law differs from other means of providing legal protection of entitlements in objects and resources, and assess whether property law is the appropriate legal response to the use and storage of bodily material.

To return to the events that gave rise to *Re: Organ Retention,* we can readily appreciate why the parents of the children sought to exercise control over the bodies on their deceased children: their relationship with the body and bodily material represented their relationship with the child. The value of this sustained relationship may provide a justification for why parents ought to be able to control or manage what happens to the bodily material of their deceased children. We can also anticipate a tension between the parents' interests in the body and the societal interest in healthcare authorities retaining and using bodily material. Although we can fit the legal dispute into Honoré's taxonomy, we can appreciate how little the taxonomy is able to explain about the way in which the law currently responds, or ought to respond, to the use and storage of bodily material. Further questions as to which branch of law is the most appropriate legal response to disputes such as *Re: Organ Retention* arise. In this particular scenario, the response provided was a legislative response in the form of the Human Tissue Act 2004.

In the remaining chapters I will argue that the appropriate response of the law to the void left by the 'no property rule' is neither a statutory response nor the wholesale application of property law.

2

The Objectification of Bodily Material

Six men were diagnosed with cancer. Since their chemotherapy treatment could damage their fertility, they were given the opportunity to produce samples of semen prior to their treatment that would then be stored by the hospital. The samples were then frozen and stored in the laboratory of the hospital. In the laboratory the amount of liquid nitrogen in the storage tank fell below the required level and the semen thawed.

On behalf of the six men, it was submitted to the Court of Appeal of England and Wales that the damage to the semen ought to constitute a personal injury.[1] The contention was that the 'intended function of the stored sperm was identical to its function when formerly inside the body, namely to fertilise a human egg', raising the question of 'why should the men's ejaculation of it make any difference?'[2] The court disagreed. To adopt the suggested approach, it was held, would be a 'fiction' where the law 'must deal in realities'.[3] Instead, the court reasoned that since the men 'generated and ejaculated the sperm', and since the 'sole object of their ejaculation of the sperm was that, in certain events, it might later be used for their benefit', the men had the right to use the semen.[4] Their right to use the semen, which was also viewed as a property right, was infringed by the hospital's failure to store the semen adequately.

It is easy to agree with the court that the six men were entitled to use their stored semen and that the men's entitlements were infringed by a failure on the part of the health authority. It is more difficult to identify a general basis for why, and under what circumstances, a progenitor ought to be able to possess, control, or use their bodily material. The difficulty stems from the status of bodily material as *separated* from the body. Items of separated bodily material are objects; they are things that are physically separate from

[1] *Yearworth v North Bristol NHS Trust* [2009] EWCA Civ 37, [18]–[24].
[2] *Yearworth v North Bristol NHS* (n 1) [19(e)].
[3] *Yearworth v North Bristol NHS* (n 1) [23].
[4] *Yearworth v North Bristol NHS* (n 1) [45(f)].

the person that can be possessed, used, transferred, destroyed, and so on. Yet, unlike other objects, items of bodily material were once part of a person. As part of the person, the bodily material was governed by the person's 'self-ownership'. The physical separation of bodily material from the person changes, to some extent, the person's relationship with the bodily material. It is this shift, from the self-ownership of the body to the ownership of bodily material, that requires investigation.

The purpose of this chapter is to identify when, and under what circumstances, a person ought to be able to possess, control, or use separated bodily material. We cannot adequately engage in this task without also touching upon more philosophical questions about the relationship between a person, their body, and their separated bodily material. Just as 'what is permissible to do with the body is partly determined by what we take the human body to be and how it is related to our own being',[5] the interests that we have in our bodies and bodily material is partly determined by what we take the human body to be and how it is related to our subjectivity or 'being'. Understanding this relationship is ultimately necessary to fill the void in the law left by the 'no property rule'. It is necessary because it helps us understand the type of ownership relationships that arise and require legal protection.

In this chapter I will explain how the physical separation of bodily material from the person changes the relationship between the person and the bodily material. I will then assess three possible bases for explaining why a person ought to be able to possess, control, or use their separated bodily material: 'prior embodiment', 'the application of work or skill', and 'the nexus of two relations between the will and object'. These justifications attempt to forge a connection between a person's (external) bodily material and a person's (inner) subjectivity.

I will then consider a fourth possible basis that differs from the earlier three insofar as it contends that a person's subjectivity is a physical subjectivity. The implication of this contention is that the physical separation of bodily material does not in all instances alienate the bodily material from the subject. This is not intended to be an exhaustive survey of possible justifications. Rather, the analysis of the four bases is intended to illustrate a philosophical contrast between conceptions of the body as arbitrarily and contingently associated with the subject and conceptions of the body as necessary and constitutive of the subject.

[5] LR Kass, 'Organs for Sale? Propriety, Property, and the Price of Progress' (1992) 107 *Pub Int* 65, 70.

This contrast between two views of the body has a number of important implications, the first of which concerns how we calibrate the relative strength of pre-social and social interests in the possession and use of bodily material. If a person's body is only contingently associated with them, then a progenitor's interest in their separated bodily material has (all other things being equal) no greater priority than any other competing interests. If, however, a person's body is constitutive of them, then a progenitor's interest in their bodily material may have priority over competing social interests.

I. Ownership and Self-Ownership

Let us start with a point of consensus. The relationship between a person and their body can be described as a relationship of 'self-ownership'. That is, with regards to their body, we can posit 'that each person enjoys, over himself and his powers, full and exclusive rights of control and use'.[6] The expression 'self-ownership' thereby describes a functional relationship of use and control by a person over their body. It confers, according to Gerald Cohen, 'the fullest right a person (logically) can have over herself provided that each other person also has just such a right'.[7] As Ngaire Naffine explains, the expression 'self-owners' has become 'a sort of legal shorthand, a rhetorical device, which serves to accentuate the fullness of the rights enjoyed by persons in relation to themselves'.[8] Note that the principle of self-ownership does not, as I explain below (at pages 61–3), presuppose a separation between the owner and the owned thing.

a. Subjectivity and Self-Ownership

Although there are different explanations for why we ought to respect a person's rights of use and control over their body, the different explanations stem from the close connection between a person's *subjectivity* (the properties or attributes of a person that grounds duties towards them) and their body. The common premise is that we have exclusive control over our bodies because the body mediates the interaction between our subjectivity and the world. As

[6] GA Cohen, 'Self-Ownership, Freedom and Equality' in N Bunnin and J Yu, (eds) *The Blackwell Dictionary of Western Philosophy* (Blackwell Publishing, 2004), 630.

[7] GA Cohen, *Self-Ownership, Freedom and Equality* (Cambridge University Press, Cambridge 1995) 213.

[8] N Naffine, 'The Structure of Self-Ownership: Or the Self Possessed Man and the Woman Possessed' (1998) 25 *JL & Soc'y* 193, 194.

Immanuel Kant explains, 'our life is entirely conditioned by our body, so that we cannot conceive of a life not mediated by the body and we cannot make use of our freedom except through the body'.[9]

Let us briefly consider two possible normative bases of the principle of self-ownership. We may, for instance, consider that each person attracts moral duties towards them because each person is autonomous. That is, each person has the capacity to reason, to formulate rationally motivated rules and judgments, and then execute 'rationally motivated moral action'.[10] If so, then rights of use and control over our bodies become necessary for the exercise of autonomy. It will only be possible to align our 'rationally motivated rules' with our 'rationally motivated actions' if each person enjoys, over himself and his powers, full and exclusive rights of control and use. Simply put, our self-governance requires, or presupposes, self-ownership.

Hence, it is possible to explain our respect for another person's rights of self-ownership with reference to his or her autonomy. If we respect another person's agency and capacity to reason, then we ought to respect the exercise of that autonomy. The key premise is that the exercise of autonomy will—first and foremost—be exercised through the body. To do otherwise, and use the body of another person without their consent, is to bypass their autonomy. Which is why, for Robert Nozick, the rights of self-ownership 'reflect the underlying Kantian principle that individuals are ends and merely means'.[11]

Alternatively, we may consider that each person has value because of the states that each person is capable of experiencing, such as states of happiness, pleasure, preference satisfaction, or well-being. The body mediates the interaction between these states and the world. The body is the primary apparatus of pain, pleasure, and well-being. Moreover, since these states are mediated by the body, 'we cannot experience from another's perspective' his or her pain, pleasure, and well-being.[12] Through each person exercising rights of use and control over their body, we enable each person to determine for themselves how best to encounter these states and maximize the experience of these states.

For example, JS Mill develops 'rules of justice' that are 'certain classes of moral rules, which concern the essentials of human well being'.[13] These

[9] I Kant, *Lectures on Ethics* (Harper & Row, 1963) 147–8.

[10] TM Powers, 'The Integrity of Body: Kantian Moral Constraints on the Physical Self' in MJ Cherry (ed), *Persons and Their Bodies: Rights, Responsibilities, Relationships* (Kluwer Academic Publishers, Dordrecht 1999) 213.

[11] R Nozick, *Anarchy, State, and Utopia* (Basic Books, 1974) 31.

[12] K Atkins, 'Autonomy and the Subjective Character of Experience' (2000) 17 *J Appl Philos* 71; see M Madden Dempsey, 'Victimless Conduct and the Volenti Maxim: How Consent Works' (2013) 7 *Crim L & Phil* 11.

[13] JS Mill, *Utilitarianism* in J Robson (ed), *The Collected Works of John Stuart Mill, Volume 10—Essays on Ethics, Religion and Society* (Toronto University Press, Toronto 1969) 250–1; JS

rules of justice thus identify and protect what Mill considers to be the most central elements of well-being or utility.[14] The most basic of these rules are the 'right to security' and the 'right to liberty'. Mill describes liberty as 'framing the plan of our life to suit our own character; of doing as we like, subject to such consequences as may follow: without impediment from our fellow-creatures'.[15] This basic right of liberty promotes well-being or utility by ensuring that, for an individual, 'over himself, over his own body and mind, the individual is sovereign'.[16] The principle of self-ownership, for Mill, reflects the basic right of liberty.

b. Separation and Self-Ownership

There are, therefore, sound bases for the principle that each person ought to have exclusive rights of use and control over their body. However, these bases of the principle of self-ownership do not naturally extend to provide an explanation for why a person ought to be able to use or control his or her own separated bodily material. When a person dies, or when a person has bodily material separated from the person, the corpse or bodily material no longer mediates the interaction between the subject and the world in the same way that the (living and attached) body does.

Hence, for some commentators, 'the notion of interests surviving death is incoherent, as there is no one who can be harmed at the point that any wrongful setback of interests occurs'.[17] John Harris argues that only 'relatively weak' and non 'person-affecting' interests persist in a person's deceased body.[18] According to Harris, 'respect for persons has two distinct dimensions: 1. respect for autonomy; and 2. concern for welfare'.[19] These

Mill, *On Liberty* in J Robson (ed), *The Collected Works of John Stuart Mill, Volume 18—Essays on Politics and Society Part I* (Toronto University Press, Toronto 1985) 27 in W Donner, 'A Millian Perspective on the Relationship between Persons and Their Bodies' in MJ Cherry (ed), *Persons and Their Bodies: Rights, Responsibilities, Relationships* (Kluwer Academic Publishers, Dordrecht 1999) 61.

[14] JS Mill *On Liberty* in J Robson (ed), *The Collected Works of John Stuart Mill, Volume 18—Essays on Politics and Society Part I* (Toronto University Press, Toronto 1985) 261 in CC Gauthier, 'Philosophical Foundations of Respect for Autonomy' (1993) 3 *Kennedy Inst Ethics J* 22, 25: '[O]ne of the principle ingredients of human happiness, and quite the chief ingredient of individual and social progress'.

[15] JS Mill, *On Liberty* 226 in CC Gauthier, 'Philosophical Foundations of Respect for Autonomy' (n 13) 25.

[16] JS Mill, *On Liberty* 224 in W Donner, 'A Millian Perspective' (n 13) 62.

[17] E Partridge, 'Posthumous Interests and Posthumous Respect' (1981) 91 *Ethics* 243; D Price, 'Property, Harm and the Corpse' in B Brooks-Gordon (ed), *Death Rites and Rights* (Hart Publishing, Oxford 2007) 199–217.

[18] J Harris, 'Law and Regulation of Retained Organs: The Ethical Issues' (2002) 22 LS 527, 548.

[19] J Harris, 'Law and Regulation of Retained Organs' (n 18) 530.

'dimensions' mirror the examples provided for the normative foundations to the above account of self-ownership. Harris suggests that none of these interests persists after the death of the individual because the bases of the 'respect for persons' are no longer applicable:

> autonomy involves the capacity to makes choices, it involves acts of the will, and the dead have no capacities—they have no will, no preferences, wants nor desires, the dead cannot be autonomous and so cannot have their autonomy violated.[20]

The dead body is no longer that which 'executes rationally motivated action' nor is it the apparatus for the experience of pleasure, happiness, or preference satisfaction. Since it is empty of these attributes, the dead body appears to be a mere object.

The same is true for separated bodily material. There is no longer an immediate connection between a subject's will, preferences, or rationally motivated judgments, and the item of separated bodily material. Pain and pleasure is no longer felt through the bodily material nor is there such an immediate and necessary connection between the bodily material and the well-being or preference satisfaction of the person. The physical separation of bodily material from the person would appear to coincide with a significant degree of separation of the bodily material from a person's autonomy or welfare, the effect of which is to place the bodily material outside the sphere of self-ownership.

Hence, the reasons for why we owe duties towards others with regards to their bodies cannot be simply extended to ground the duties towards others with regards to their bodily material or posthumous bodies. The values that ground our respect for another's self-ownership of his or her body are values that are not present, to the same extent, in bodily material.

The physical separation of bodily material, at the very least, dilutes these values. For instance, recall the plight of Ms Evans in *Evans v Amicus Healthcare.* Had the embryo been implanted in her, or had she conceived naturally, then her right to bodily integrity (that is predicated on her 'self-ownership') would have enabled her to carry the embryo, and become a genetic mother, despite the wishes of Mr Johnson. Her right to bodily integrity would outweigh Mr Johnson's right not to be a genetic parent. Instead, the embryo was separated and, as Mary Ford explains, a different 'kaleidoscope of interest and rights' emerged.[21] Ms Evans' interest in the embryo could only be based on her right to respect for her private and family life, as was Mr Johnson's interest in the

[20] J Harris, 'Law and Regulation of Retained Organs' (n 18) 531.

[21] M Ford, '*Evans v United Kingdom*: What Implications for the Jurisprudence of Pregnancy?' (2008) 8 *HRL Rev* 171, 181.

embryo. The separation of the embryo enabled Mr Johnson's interest in the embryo to be 'full-bodied and capable of defeating those of' Ms Evans.[22]

c. Prior Embodiment

Despite the separation of bodily material from a person, it is nonetheless intuitive that the material ought to still *belong* to the progenitor (the person to whom material was attached). The bodily material *was* part of the person. The person was, in other words, the first to acquire possession of the material. This 'prior embodiment' or 'original acquisition' is considered by some to be sufficient to ground a person's entitlements to use and control his or her material. Rohan Hardcastle, for instance, argues that 'property rights created on detachment represent a natural extension of the right to bodily integrity'.[23] William Boulier argues that if society views the body of an individual as their property, 'logic and justice' requires that property rights are recognized in the separated body parts.[24] Muireann Quigley suggests that our 'self-ownership' 'raises, at least, the strong presumption that we own our body's separated parts and products',[25] whilst J Kenyon Mason and Graeme Laurie suggest that 'if a tissue sample could be property' then it is 'more attractive from a moral and ethical standpoint to adopt the position ... that the original owner was the individual from whom it was taken.'[26]

[22] M Ford, '*Evans v United Kingdom*' (n 21) 183.

[23] R Hardcastle, *Law and the Human Body: Property Rights, Ownership and Control* (Hart Publishing, Oxford 2009) 147: 'Recognising that property rights are created on detachment represents a natural extension of the right to bodily integrity. Before separation, biological materials remain part of the human body. The law provides protection for the human body, and such protection is enforceable generally against other members of society. It would seem inconsistent if the act of detachment changed biological materials from material fully protected by law into material receiving no legal protection whatsoever. The law can provide protection to physically detached biological materials by employing the "bodily integrity principle". To do this, it is necessary to create property rights.'

[24] W Boulier, 'Sperm, Spleen and Other Valuables: The Need to Recognise Property Rights in Human Body Parts' (1995) 23 *Hofstra L Rev* 693, 717: 'At some level, it makes perfect sense to call one's body one's property, and on this ground alone commentators have suggested that the law make this recognition. If this is indeed the general view in society, logic and justice seem to argue for the recognition of a property right.'

[25] M Quigley, 'Property and the Body: Applying Honoré' (2007) 33 *J Med Ethics* 631–4: 'We can each claim to have "full liberal ownership of our bodies", and to have property rights in our own bodies. This also raises, at least, the strong presumption that we own our body's separated parts and products.'

[26] JK Mason and GT Laurie, 'Consent or Property? Dealing with Body Parts in the Shadow of Bristol and Alder Hey' (2001) 64 *Mod L Rev* 710, 724: Mason and Laurie refer to the Medical Research Council, *Working Group on Human Tissue and Biological Samples for Use in Research: Report of the Medical Research Council Working Group to Develop Operational and Ethical Guidelines* (1999) [2.2.1.]: 'it was more practical and more attractive from a moral and ethical

We have seen that a person ought to be viewed as having exclusive rights of control and use of his or her body because of the immediacy of the connection between a person's subjectivity and his or her body. The difficulty is that the normative bases of the principle of self-ownership cannot simply be substituted to provide the normative basis of entitlements in separated bodily material. There is no longer a necessary and immediate connection between a person's subjectivity and his or her bodily material. The prior embodiment principle cannot therefore be seen as a natural extension of the principle of self-ownership or the right to bodily integrity.

The problem with the prior embodiment justification is that it does not explain why prior acquisition or possession is a past act or event that can, in the present, impose on others duties to refrain from using the resource.[27] There is no reason to presume that someone's prior relationship with bodily material relates, or is connected, to his or her current assertion of rights in the bodily material. Something further is required in order to show why the progenitor's prior embodiment is sufficient to explain why the progenitor ought to be able to—in the present—control or use the bodily material. The self-ownership of the body imposes duties out of respect for autonomy or respect for welfare. Yet, this on-going duty to respect the entitlements of the first to acquire the bodily material is, at best, an unfamiliar moral duty.[28]

For example, in *Moore v Regents* Dr Golde owed Mr Moore legal duties with regards to his body, including his spleen cells.[29] Golde was held to be under a duty to disclose to Moore his research interest in Moore's spleen in order for Moore's consent to surgery and subsequent tests to be informed consent.[30] This reflects Golde's duties to Moore as a self-owner; Moore's decisions with regards to his body (his preferences or rationally motivated decisions) ought to be informed decisions. The spleen cells, as separated bodily material, were originally—but are no longer—part of the sphere of self-ownership. All other things being equal, Golde does not undermine Moore's autonomy or

standpoint to adopt the position that, if a tissue sample could be property, the original owner was the individual from whom it was taken.'

[27] J Simmons, 'Original-Acquisition Justifications of Private Property' (1994) 11 *Soc Phil & Pol'y* 63 summarises the contention advanced by a number of sceptics: '[Original Acquisition] arguments normally entail that persons can, by their acts of acquisition, deliberately create for others universal moral duties of forbearance and noninterference with respect to holdings of possibly scarce resources. Such a power (to create significant moral burdens for others at will) is "radically unfamiliar" and "repugnant" to us, and it is therefore a power of which we should be highly suspicious. Needless to say, we should be equally suspicious of any argument which purports to justify such a power for individuals.'

[28] J Waldron, *The Right to Private Property* (Clarendon Press, Oxford 1988) 265–71.

[29] *Moore v Regents of the University of California* 793 P 2d 478 (Cal 1990).

[30] *Moore v Regents of the University of California* (n 29) 128–34.

set back his interests or preferences by using the separated spleen cells. As the Supreme Court of California held, Moore was wronged by Golde, but only because Moore was not afforded the right to be fully informed about intrusions into his body, and not because a subsequent person used bodily material that was first possessed, or originally acquired by, Moore.

The underlying problem with the prior embodiment justification is that it fails to explain why the mere fact that a person possessed or embodied the resource *first* is able to ground a duty on another person not to use the bodily material once it is no longer possessed or embodied. The prior embodiment justification does not draw upon the consequences of the progenitor being able to control his or her separated bodily material, nor does the prior embodiment justification rely upon a continuing connection between a person's subjectivity and his or her body that we ought to respect. Rather, the mere fact of prior embodiment is presumed to provide sufficient reason for the original acquirer to obtain entitlements in the resource ahead of any other potential entitlement-holders.

d. The Application of Work or Skill

Even if our self-ownership is limited to the ownership of our bodies, it follows from our rights of use and control over our bodies that we *own our labour*. That is, we exercise our autonomy, or satisfy our preferences, through the use of our bodies towards the performance of a task. The performance of the task is the exercise of *our* labour. Given the starting point (i) that we *own our labour*, we may add the additional premise that (ii) a person who labours on an object mixes that labour with that object, to conclude that (iii) the object which has been laboured on contains something which the labourer owns.[31] As Jeremy Waldron explains, 'to take the object out of the labourer's control without his consent would violate' a person's ownership of their labour that is mixed in the object.[32] I will argue here, that as a pre-social justification for why a person ought to be able to exercise entitlements in bodily material, the Lockean 'work or skill' principle provides an incomplete

[31] J Waldron in 'Two Worries about Mixing One's Labour' (1983) 33 *Phil Q* 39 reconstructs Locke's argument from *Two Treatises of Government* as follows: '(1) A person who labours on an object mixes his labour with that object. (2) But that person owns the labour which he mixes with the object. (3) So the object which has been laboured on contains something which the labourer owns. (4) To take the object out of the labourer's control without his consent would violate the entitlement mentioned in (2) and (3). (5) Therefore, no one may take the object out of the labourer's control without his consent. (6) But this amounts to an entitlement in the labourer over the object. (7) Thus, a person who has laboured on an object is entitled to that object.'

[32] J Waldron in 'Two Worries about Mixing One's Labour' (n 31) 39.

connection between the characteristics and attributes of the person and the bodily material. In terms of the 'work or skill rule' as an accepted legal basis for the recognition of the right to possess bodily material as an item of property, it is best understood as a *social* justification for the possession, use, or control of bodily material.

To start, we need not accept the intermediate premise; that (ii) a person who labours on an object mixes that labour with that object. The labourer (or the applier of work or skill) is not being mixed with the object. For instance, when human tissue is set in a resin slide, semen is frozen in liquid nitrogen, or even when genes are isolated through a polymerase chain reaction, one object (the bodily material) is mixed with another object (the resin, the nitrogen, the enzyme), and it is the labour that is doing the mixing. At best, when work or skill is applied to an object, the appliers of work or skill are 'mixing their mixing'.[33]

That said, the application of work and skill may, through the intentional acts of the labourer, change the character of the object. In such instances, it can be said that the work or skill has produced a quality in the object that connects the subject with the object. Because of the intentional acts of the person, the object acquires a new quality that is attributable to the autonomy or preferences of the subject. The subject is, in a limited way, mixed with the object. This explanation only goes some of the way to link the attributes and characteristics of the person to the object. It could be said that the gap between subject and object is partially bridged by appliers of work or skill introducing a new attribute to the object *through* them. The attributes of the object cannot be accounted for without reference to the persons who applied *their* skill and labour. The first exception to the 'no property rule' was developed along these lines. Recall that the High Court of Australia in *Doodeward v Spence*[34] held that where a person had:

> by the lawful exercise of work or skill so dealt with a corpse in his lawful possession that it had acquired some attributes differentiating it from a mere corpse, that person acquired a right to possession of the corpse or part.[35]

Because of the application of 'work or skill' sufficient for the foetal twins to acquire some new attribute, the preserved foetal twins were items of property, and the claimant could rely upon actions in conversion and detinue to retrieve possession of the bodily material.

[33] J Waldron, 'Two Worries about Mixing One's Labour' (n 31) 41.
[34] *Doodeward v Spence* (1908) 6 CLR 406. [35] *Doodeward v Spence* (n 34) 414.

Yet, this account of ownership based upon the single causal link between actions of the person and the attributes of the object is over-inclusive. We frequently introduce new attributes in objects through our intentional actions without our subjectivity (our autonomy, will, or preferences) necessarily being invested or manifest in the object. Take Nozick's well-known objection to the mixing argument:

> why isn't mixing what I own with what I don't own a way of losing what I own rather than gaining what I don't? If I own a can of tomato juice and spill it in the sea so that its molecules mingle evenly throughout the sea, so I thereby come to own the sea, or have I foolishly dissipated my tomato juice?[36]

Creating a new attribute in the object is, by itself, insufficient to explain why the object is brought within the sphere of the subject instead of the work or skill being 'dissipated' into the external world.

Even if we accept that the application of work and skill can create a link between the subject and object, a fundamental assumption is that the link between subject and object is sustained. The single causal connection between subject and object is unable to explain why our duties to respect a person's liberty over their actions (in the past) extend to ground a duty to respect a person's ownership in an object (now). As Waldron argues, 'ownership in this sense has no relevance *now* to labour that has been freely expended in the past'.[37] The contention is that the mere (past) use of labour to mix objects is insufficient to justify (now) entitlements in one of the objects.

The 'work or skill' exception to the no property rule need not rely on a connection between the attributes and characteristics of the person and the item of bodily material in order for the exception to have normative force. This is because the exception is best explained as providing a social basis for the exercise of the right to possess, or even profit from, bodily material. The 'application of work or skill' is an activity that the law seeks to incentivize or encourage. For instance, the preservation of body parts for medical or scientific examination is socially valuable, and in order to encourage the activity, the law protects the 'labourers' entitlement to possess the body parts. The Court of Appeal of England and Wales affirmed the *Doodeward* exception to the 'no property rule' in *Dobson v North Tyneside Health Authority* and *R v Kelly*. In *Dobson v North Tyneside Health Authority*, it was held that minimal work or skill that was applied to the bodily material (the preservation of the brain in fluid) was insufficient for the 'application of skill' exception.[38]

[36] R Nozick, *Anarchy, State, and Utopia* (n 11) 175.

[37] J Waldron, 'Two Worries about Mixing One's Labour' (n 31) 44.

[38] *Dobson and another v North Tyneside Health Authority and another* [1996] 4 All ER 474, 479.

Whereas in *R v Kelly*, since the body parts 'had undergone a process or application of human skill designed to preserve it for medical or scientific examination, it acquired a value and became property for the purposes of the Theft Act 1968'.[39] Recall also that s 32(9)(c) of the Human Tissue Act 2004 provides that 'controlled material' that 'is the subject of property because of an application of human skill' is exempted from the prohibition of commercial dealings in human tissue.[40] The additional requirements that the work or skill be of a sufficient degree, or to result in a new attribute in the bodily material, are best interpreted as a threshold for the work or application of skill that is sufficiently valuable to be rewarded with the entitlement to possess, or profit from, the bodily material.

e. The Nexus of Dual Relations

The work or skill exception goes some, but not all, of the way to bridging the gap between object and subject. It provides a single causal connection between the person (subject) and the bodily material (object) on the basis that the person introduced a new attribute to the object *through* the person. I will suggest here a Hegelian account of ownership is able to fully bridge the gap.

Georg W F Hegel describes the individual as having an 'inherently individual will' which is 'purely abstract'.[41] Hegel argues that in order to develop an individual's personhood, the individual's 'will' or subjectivity must develop a 'content of determinative ends'.[42] The will has 'content' when the will is manifest in the external world.[43] Hence, for Hegel, 'a person must translate his freedom into an external sphere in order to exist as Ideal'.[44] In other words, the will of the individual, or the content of the person's subjectivity, cannot have value in the abstract. It needs to be expressed through engagement with the external and physical world. This is an important insight. Whatever we take as the basis of respect for a person (his or her autonomy, welfare, preference satisfaction, human dignity, or human flourishing), we can either view that subjectivity as abstract and free-floating attributes or qualities, or

[39] *R v Kelly/R v Lindsay* [1998] 3 All ER 741, 750.

[40] Human Tissue Act 2004, s 32(1).

[41] AW Wood (ed) and GWF Hegel, *Elements of the Philosophy of Right* (HB Nisbet (tr)) (Cambridge University Press, Cambridge 1991) [34].

[42] GWF Hegel, *Elements of the Philosophy of Right* (n 41) [34].

[43] GWF Hegel, *Elements of the Philosophy of Right* (n 41) [34]; MJ Davis and N Naffine, *Are Persons Property? Legal Debates about Property and Personality* (Ashgate, Aldershot 2001) 4–5.

[44] GWF Hegel, *Elements of the Philosophy of Right* (n 41) [41].

as attributes or qualities that are engaged and manifest in the messy, external, and physical world.

The way in which an element of the will or a person's subjectivity is expressed in an object is described by Waldron as involving a 'conjunction of two relations between my will and the [object]'[45] or a 'nexus of dual relations between the will and object, object and will'.[46] Starting with *the first relationship* between the will and the object, Hegel suggests that an object becomes 'affected by the will' when the object acquires an attribute or characteristic, because of the use or possession of the object, 'which can only be explained in terms of the working will'.[47] However, as we know, we frequently alienate objects that have characteristics that can only be attributed to our will without a sense of alienating an element of our personhood or subjectivity. As discussed above, the mere change in an object through the actions of the subject does not necessarily pull the object into the sphere of the subject.

What is required for the object to embody or manifest our will or subjectivity is for the object to 'affect the will'. This is the *second relationship* of the nexus. An object affects the will by enabling a certain state of affairs that the individual intends, desires, or 'wills' to exist, and that would otherwise not exist without the object. It is when an object is affected by the will through acquiring a new characteristic *and* the object affects the will by enabling a new state of affairs that a person's will or subjectivity is mixed with, or invested into, the object. A Hegelian basis for entitlements in bodily material would suggest that:

(i) where the bodily material acquires an attribute that can only be explained 'in terms of the working will'; and
(ii) where that attribute introduces an intended state of affairs,
(iii) entitlements in the material may arise because the will (or subjectivity) of the individual needs to be expressed through engagement with the external and physical world.

Note the blurring between object and subject in the Hegelian explanation. To the extent that we divide the world up into objects and subjects, it is unclear whether an element of the will or subject is extended out into the external sphere or an object of the external sphere has become part of the subjectivity of the person. We will return to this Hegelian blur in chapter four (at pages 128–31), and we will consider in the next section a view that

[45] J Waldron, *The Right to Private Property* (n 28) 369.
[46] J Waldron, *The Right to Private Property* (n 28) 370.
[47] J Waldron, *The Right to Private Property* (n 28) 369.

ultimately dissolves the distinction between the external sphere of objects and the inner subjectivity of a person.

f. Yearworth and the Hegelian Basis

In terms of the development of the law, this Hegelian analysis aligns with the 'new basis' set out by the Court of Appeal in *Yearworth v North Bristol NHS*. In particular, the 'dual nexus' mirrors the explanation provided by the Court of Appeal for why the claimants obtained the right to use their stored bodily material. It is because the claimants:

(i) By their bodies, they alone generated and ejaculated the sperm,

(ii) The sole object of their ejaculation was that it might later be used for their benefit…

(iii) The men had ownership of the sperm for the purposes of their present claims… namely in relation to its future use.[48]

To put the reasoning in *Yearworth v North Bristol NHS* in Hegelian language, since (i) the semen acquired an attribute 'which can only be explained in terms of the working will'[49] and (ii) the semen enables an intended state of affairs that would otherwise not exist without the object, the men own the semen since (iii) the will (for genetic parenthood) can only be manifest through the use of the semen.

This Hegelian interpretation of *Yearworth v North Bristol NHS* provides a promising approach to the recognition of entitlements in bodily material and may address some of the 'unanswered questions' following the Court of Appeal's decision. For instance, non-reproductive tissue may be approached in the same manner,[50] so that use entitlements in the tissue are retained where the purpose of the extraction is that the tissue 'might later be used for their benefit' or for the benefit of a co-participant in therapy.[51] The scope of the progenitor's rights[52] are limited to the rights necessary to attempt to achieve the intended 'state of affairs'. In *Yearworth v North Bristol NHS*, the scope of the progenitors' rights where limited to the right to management and use

[48] *Yearworth v North Bristol NHS* (n 1) [45(f)].

[49] J Waldron, *The Right to Private Property* (n 28) 369.

[50] SHE Harmon and GT Laurie, '*Yearworth v North Bristol NHS Trust:* Property, Principles, Precedents and Paradigms' (2010) 69 *CLJ* 486–7: 'How should non-reproductive tissue be approached?'

[51] SHE Harmon and GT Laurie, 'Property, Principles, Precedents and Paradigms' (n 50) 487: 'What about material taken for transplantation into another or back into oneself?'

[52] SHE Harmon and GT Laurie, 'Property, Principles, Precedents and Paradigms' (n 50) 487: 'What is the scope of originators' rights to treat with their own body and parts?'

the semen. As I will explain in the next chapter (at pages 81–8), the scope of the progenitor's rights cannot, on this basis, include the right to 'participate in bio-markets'.[53]

However, since the decision in *Yearworth v North Bristol NHS*, courts have been willing to recognize property rights in semen upon more general and less precise bases. Consider two examples. The Supreme Court of Queensland in *Bazely v Wesley Monash IVF*[54] recognized property rights in semen that the progenitor extracted and stored with a fertility clinic prior to his death. The Court held that 'both in law and in common sense, [the conclusion] must be that the straws of semen currently stored with the [fertility clinic] are property, the ownership of which vested in the deceased while alive and in his personal representatives after his death'.[55] The Court, with general reference to the decision in *Yearworth v North Bristol NHS*, the work or skill rule in *Doodeward v Spence*, and the right to possession as the corollary of the duty to inter under *Re: Gray*,[56] reasoned that where a person has lawfully applied work or skill to a human body or part such 'that it has acquired some attributes differentiating it from a mere corpse awaiting burial', the person obtains the right to possession as a property right.[57] As a result, the deceased's entitlements in his semen were property rights and the wife of the deceased was able to obtain these property rights through the application of Section 8 of the Queensland Succession Act 1981.

The Supreme Court of New South Wales, in *Re: Edwards*,[58] also afforded stored semen the legal status of property so that the widow of the deceased progenitor could obtain possession of the semen. The crucial difference between the extractions in *Re: Edwards* and the extraction in *Bazely* is that in the former case the semen was extracted from a deceased patient by court order. The reasoning of the Supreme Court of New South Wales revolved around two justificatory bases. First, the Court was satisfied, following *Re: Gray*, that the widow had a limited right to possess the bodily material as the corollary of the duty to inter.[59] Yet, the court also held that the widow had a right to the semen that 'extends beyond that which she would have as administrator' of the estate.[60] Second, the semen had the status of property because of the application of work or skill rule in extracting and storing the

[53] SHE Harmon and GT Laurie, 'Property, Principles, Precedents and Paradigms' (n 50) 487: 'What, now, is the potential for originators of tissue to participate in bio-markets?'

[54] *Bazley v Wesley Monash IVF Pty Ltd* [2010] QSC 118.

[55] *Bazely v Wesley Monash IVF* (n 54) [33].

[56] *Re: Gray* [2000] QSC 390.

[57] *Bazely v Wesley Monash IVF* (n 54) [18].

[58] *Jocelyn Edwards; Re the Estate of the Late Mark Edwards* [2011] NSWSC 478.

[59] *Re: Gray* [2000] QSC 390; *Re: Edwards* (n 58) [41]–[42].

[60] *Re: Edwards* (n 58) [90].

bodily material, and it became *her* property because the semen was extracted 'for *her* purposes'.[61]

In terms of the rule in *Re: Gray*, the right of the administrator of an estate to obtain possession of the body or bodily material of the deceased follows as a corollary of the administrator's duty to inter the body of the deceased. This right to possession ought to be circumvented by the reasons for which the right arises.[62] Yet, in *Bazely v Wesley Monash IVF* and *Re: Edwards* the widows obtained general rights of possession that were acknowledged to encompass the capacity to use the bodily material. Hence, the entitlements recognized in *Bazely v Wesley Monash IVF* and *Re: Edwards* cannot follow from the reasoning in *Re: Gray*, where it was held that:

> [A] deceased's personal representative or, where there is none, the parents or spouse, have a right to possession of the body *only for the purposes of ensuring prompt and decent disposal* [which] has ... the corollary that there is a duty not to interfere with the body or ... to violate it.[63]

Hence, the courts appear to have collapsed the distinction between the right to obtain a *limited* right to possession under the rule in *Re: Gray* (or *Dobson v North Tyneside Health Authority*) and the general right to possess *and use* the bodily material of the deceased that arises under the exceptions to the no property rule in *Yearworth v North Bristol NHS* or *Doodeward v Spence*.

Perhaps then the right for a partner to obtain the right to use the bodily material of the deceased 'extends beyond'[64] the right of a partner as the administrator of the deceased's estate. The basis for finding that the widows in *Bazely v Wesley Monash IVF* and *Re: Edwards* had the right to possess and use the stored semen appears to follow from application of the work or skill rule under *Doodeward v Spence*. Yet, the work or skill rules provides a basis for the *applier* of work or skill to obtain possession and use entitlements in bodily material. Since the widows in *Bazely v Wesley Monash IVF* and *Re: Edwards* were neither the appliers of the work or skill nor the progenitors of the bodily material, the entitlements in the bodily material must be either be lawfully *transferred* to them from the applier of work or skill, or the work or skill can be applied *vicariously* 'for their purposes',[65] in order for them to obtain the entitlements in the bodily material. The transfer of the semen from the applier of work or skill can account for the circumstances in

[61] *Re: Edwards* (n 58) [91] [emphasis added].

[62] *Re: Gray* (n 59); *Dobson and another v North Tyneside Health* (n 38).

[63] *Re: Gray* (n 59) 390 [emphasis added].

[64] *Re: Edwards* (n 58) [90].

[65] *Re: Edwards* (n 58) [91]; L Skene, 'Proprietary Interests in Human Bodily Material: Yearworth, Recent Australian Cases on Stored Semen and Their Implications' (2012) 20 *Med L Rev* 227, 237.

Bazely v Wesley Monash IVF since the progenitor obtained property rights in his semen that were then transferable to his widow (through Section 8 of the Queensland Succession Act 1981). However, it was the fertility clinic that applied work or skill to extract the semen in *Re: Edwards* and the possible transfer of rights from the clinic to the widow confronted a statutory prohibition (s 21 of Assisted Reproductive Technology Act 2007).

Alternatively, to suggest that the widow in *Re: Edwards* ought to obtain the right to use the bodily material because it was removed 'for her purposes',[66] either as an attempt to fit the circumstances into the reasoning in *Yearworth v North Bristol NHS* or to claim that the application of work or skill was applied 'for her benefit', is also problematic. The semen was removed following a court order as an interim measure. To say that it was 'for her purposes', rather than removed for the purposes of a later judicial determination, is to presume that she had the right to the semen when it was extracted despite the fact that the semen was extracted in order for the court to determine whether the widow had a right to the semen. This interpretation of *Re: Edwards* lends itself to a self-fulfilling prophecy since it pre-supposes that the purposes for which the bodily material is removed for are purposes that the claimant is legally entitled to pursue. More generally, it is otherwise circular to suggest that a person can have bodily material removed from another person, and then point to the fact it was removed 'for their purposes' in order to validate the purposes for which they intend to use the bodily material. To broaden the reasoning in *Yearworth v North Bristol NHS* and formulate 'a principle that one can lawfully remove tissue from a corpse, for one's own purposes, provided the corpse is in one's lawful possession' is, as Loane Skene argues, 'surely too wide'.[67]

Note, however, that the finding in *Bazely v Wesley Monash IVF* can be reconstructed consistently with the Hegelian basis of ownership. Since the deceased:

(i) by his body, generated and ejaculated the sperm and thereby acquired an attribute 'which can only be explained in terms of the working will',[68]

[66] *Re: Edwards* (n 58) [91]; L Skene, 'Proprietary Interests in Human Bodily Material' (n 65) 237: 'if a court authorises the removal of bodily material from a person for later use by another person, that may provide the basis for the tissue becoming property by work or skill performed for the purposes of a particular person by an agent.'

[67] L Skene, 'Proprietary Interests in Human Bodily Material' (n 65) 237: Skene provides a good example of why we should be wary of conflating possession with use: 'Consider a man who suspects he is not the biological father of a child who has just died and takes a blood sample for DNA analysis. Surely a court would not say that, even if he is legally entitled to possession of the body for burial, he is also legally entitled to take a sample and test it for his own purposes.'

[68] J Waldron, *The Right to Private Property* (n 28) 369.

(ii) the object of which was to use the sperm for his (and his partner's) benefit, and thereby enables an intended state of affairs that would otherwise not exist without the object,

the deceased obtained (iii) the right to use the semen for the purposes of the intended state of affairs. Rather than relying on the law of succession, perhaps the appropriate extension is to add (iv) that the partner, as a co-participant in the reproductive therapy (or intended state of affairs), obtains a reversionary interest in the semen that vests in the co-participant in circumstances such as the death of the progenitor. The reasoning in *Re: Edwards*, in comparison, cannot be reconstructed consistently with either the exceptions identified in *Doodeward v Spence, Yearworth v North Bristol NHS*, or the rule in *Re: Gray*.

As we have encountered in this section, the separation of bodily material from the person can be interpreted as creating a gap between the person (the subject) and the bodily material (the object). Since the reasons for why we respect the rights of use and control of a person over his or her body cannot simply extend over this gap, branches of both the law and philosophy have sought techniques for bridging the gap between a subject's self-ownership and the ownership of objects. The Lockean requirement of the application of work or skill is unable to bridge the gap but is able to provide a social justification for the exercise of entitlements in bodily material that has obtained a new attribute. In contrast, the Hegelian requirement of the dual nexus between will and object is able to bridge the gap between object and subject. We have seen how the reasoning in *Yearworth v North Bristol NHS* aligns with the Hegelian justification whilst the subsequent applications of the decision in *Yearworth v North Bristol NHS* have departed from the sound justificatory basis contained in the judgment. We shall now consider whether the distinction between object and subject is best interpreted as a 'gap'.

II. Self-Ownership and Physical Subjectivity

We identify a gap between the subject (the progenitor) and the object (the bodily material), and then search for ways to bridge the gap, because we have become used to thinking of ourselves as having two levels of existence. That is, a personal level of existence, which is characterized by subjective feelings, mental representations and perceptions, as well as having a sub-personal level of existence, on which level we regard ourselves as made up of physiological systems that carry out their functions largely unbeknown to us.[69] We presume that our thoughts and feelings are 'us' and the body is a physiological system in which we are encased. In short, 'we regard perception as subjective, and the body as objective'.[70] Or at best, the body is 'a thing "inserted" between the rest of the world and the "subjective sphere"'.[71]

Hence, the discussion so far has proceeded on a variety of dualism between the *subject*, as the will, rationality, consciousness of the person, and the *body*, as a type of physical entity in the world of objects.[72] The principle of self-ownership has so far been seen as a relationship of use and control of the body by the subject that can be explained with reference to the close causal connection between the body and the subject's capacity for rationality, utility, or well-being. The Lockean 'work or skill rule', and Hegelian 'nexus of dual relations', have attempted to explain how a person's labour or a person's 'abstract will' can be mixed in, or extended into, objects. The underlying assumption here is that the subject is an inner entity that uses the body-object to work with, and extend itself into, the physical-material world of objects.

a. A Physical Subjectivity

In contrast to this underlying presumption, Maurice Merleau-Ponty's contention is that our 'subjectivity is physical'.[73] The suggestion here is that 'body and experience are not just causally connected, but identical'.[74] Our personal existence of subjective feelings and perceptions are not distinct from our sub-personal existence of physiological systems. The 'structures and capacities of the body' *are our* perceptions.[75] Merleau-Ponty contends that my body 'is my point of view on the world'.[76] In other words, our

[69] M Merleau-Ponty, *Phenomenology of Perception* (C Smith (tr)), (Routledge, London 2003) 84: 'Where it was desired to insert the organism in the universe of objects... it was necessary to translate the functioning of the body into the language of in-itself and discover, *beneath behaviour, the linear dependence of stimulus and receptor, receptor* Empfinder.' [emphasis added].

[70] T Carman, *Merleau-Ponty* (The Routledge Philosophers, Routledge, 2008) 83.

[71] M Biemel (ed), E Husserl, *Ideen zu einer reinen Phänomenologie und phänomenologischen Philosophie. Zweites Buch: Phänomenologische Untersuchungen zur Konstitution. Husserliana IV* (Martinus Nijhoff, The Hague 1952) 161; T Carman, 'The Body in Husserl and Merleau-Ponty' (1999) 27(2) *Phil Topics* 205, 206.

[72] See S Priest, *Merleau-Ponty* (Arguments of the Philosophers Series, Routledge, London 1998) 29–30.

[73] S Priest, *Merleau-Ponty* (Arguments of the Philosophers Series, Routledge, London 1998) 57: 'Merleau-Ponty's originality lies in the idea that subjectivity is physical... that I am a subjective object or a physical subject.'

[74] T Carman, *Merleau-Ponty* (n 70) 81.

[75] T Carman, *Merleau-Ponty* (n 70) 81.

[76] M Merleau-Ponty, *Phenomenology of Perception* (n 69) 81.

bodies are the site or location of where our subjectivity engages with the world, and there remains no 'inner' space to which the subject can retreat.[77]

Moreover, since the body 'represents my particular point of view *on* the world' it is a point of view 'on which ... I can take no point of view'.[78] As Merleau-Ponty explains, 'as for my body, I do not observe it itself: to be able to do so, I would need the use of a second body, which would not itself be observable'.[79] Merleau-Ponty then concludes that 'what prevents [the body] ever being an object, ever being "completely constituted," is that it is that by which there are objects'.[80] The body, therefore, is complicated: it cannot be distinguished from the subject nor does it equate to the other objects that we encounter, since the body is the way in which we encounter objects.

Allow me to further outline in this section some aspects of Merleau-Ponty's view of the body and extend these views to formulate a conception of how our subjectivity relates to our bodies and items of bodily material. I will suggest that the body can be understood as an object that is *for-itself* and *for-others* or as an object *in-itself*. This tripartite account of the body becomes fundamental to the discussion of the body and bodily material through the remainder of this book.

b. The Body For-Itself

If the body is the location of the subject, it follows that the body is an object that 'cannot be understood as an object merely occurring in the world'.[81] Rather, the body is an object *for-itself* ('*pour soi*' or '*qua* oneself').[82] The body is the way in which a person exists in the world, experiences the world, connects with others, and engages in tasks and projects. As Merleau-Ponty explains:

> The body is the vehicle of being in the world, and having a body is, for a living creature, to be intervolved in a definite environment, to identify oneself with certain projects and be continually committed to them.[83]

[77] M Merleau-Ponty, *Phenomenology of Perception* (n 69) 11: '... there is no inner man, man is in the world, and only in the world does he know himself.'

[78] SK Toombs, 'What Does it Mean to Be Somebody?: Phenomenological Reflections and Ethical Quandaries' in MJ Cherry (ed) *Persons and Their Bodies: Rights, Responsibilities, Relationships* (Kluwer Academic Publishers, Dordrecht 1999) 79.

[79] M Merleau-Ponty, *Phenomenology of Perception* (n 69) 91.

[80] M Merleau-Ponty, *Phenomenology of Perception* (n 69) 92.

[81] T Carman, *Merleau-Ponty* (n 70) 102.

[82] S Priest, *Merleau-Ponty* (n 73) 67: 'For Merleau-Ponty the body *qua* oneself exhibits the phenomenological-ontological property of being "pour soi" "for-itself".'

[83] M Merleau-Ponty, *Phenomenology of Perception* (n 69) 94.

The body, as the means of existing in, and engaging with, the world, represents an 'attitude directed towards a certain existing or possible task'.[84] We can interpret a 'task' as something akin to the Hegelian idea of the 'abstract will that is manifest in the external sphere'. A 'task' represents the qualities or attributes of the person that are engaged and manifest in the messy, external, and physical world. The body is *for-itself* since it is the way the subject forms 'intentional ties' with the world, undertakes 'tasks', and identifies itself with certain projects. Yet, unlike the Hegelian subject that must 'extend itself into the external sphere', Merleau-Ponty would suggest that the subject already finds him- or herself extended into the external sphere, *as* the sum of 'intentional ties' with the external sphere.[85] The subject *is* the sum of connections or intentional ties with the world.

Here we can identify a first-order aspect of our subjectivity. Given that a person's subjectivity (however constituted) cannot exist in the abstract and is located in the external and physical world, the way in which a component of subjectivity (a plan, a goal, a preference, an experience) manifests itself is, prior to anything else, through the body. The body or bodily material that is *for-itself* is a first-order aspect of our being or subjectivity.

c. The Body For-Itself and an Object In-Itself

Although the body is part of the physical-material world (is an object), the body is not a mere object. 'The definition of the object' for Merleau-Ponty is 'that it exists parte extra parte [parts outside of parts]'.[86] There exists only 'mechanical relationships' between objects that interact through direct causal contact.[87] The body-object is not 'instrumental' in the same way that mere-objects are instrumental. As Jean-Paul Sartre explains, 'I am not in relation to my hand in the same utilizing attitude as I am in relation to the pen; I am my hand ... we do not use this instrument for we are it.'[88]

The contention here is that the body-object is orientated towards tasks (understood in terms of the interwovenness of the subject with the world). The body is always experienced as 'in the midst of environing things, in this or that action, posited or positioned relative to some task' where the task is figuratively or literally 'at hand'.[89] A mere-object (or an objects *in-itself*

[84] M Merleau-Ponty, *Phenomenology of Perception* (n 69) 100; SK Toombs, 'What Does it Mean to Be Somebody' (n 78) 81.

[85] S Priest, *Merleau-Ponty* (n 73) 68: 'Pre-philosophically a person experiences him or herself as a whole acting person embedded in the world.'

[86] M Merleau-Ponty, *Phenomenology of Perception* (n 69) 84; T Carman, *Merleau-Ponty* (n 70) 83.

[87] M Merleau-Ponty, *Phenomenology of Perception* (n 69) 84; T Carman, *Merleau-Ponty* (n 70) 83.

[88] JP Sartre, *Being and Nothingness: A Phenomenological Essay on Ontology* (H Barnes (tr)), (Philosophical Library, New York 1956) 426–7.

[89] RM Zaner, *The Context of Self: A Phenomenological Inquiry Using Medicine as a Clue* (Ohio University Press, Athens 1981) 97.

or *en soi*),[90] in contrast, is only oriented in actions where there exists a causal-mechanical relation between the body-object and the mere-object (or as between two mere objects). The hand, therefore, is engaged in the task of extending thoughts and representations into observable strokes of ink; the pen is engaged in a series of causal connections from the application of downwards pressure by the hand to the friction provided by the paper. The mechanical act of writing is a causal mechanical action, whereas the intentional act of writing is a 'task'.

d. The Self-Ascribed Body

Given this distinction between the body that is *for-itself* and an object that is *in-itself*, what body counts as the body (that is *for-itself*) is then understood in terms of responsiveness to tasks. The body is 'polarised by its tasks, it *exists towards* them, it gathers itself up to reach its goal'.[91] The body-object is therefore the unified response to a series of tasks: the demands of coping with the world, our ties or connections with the world, and our projects in the world. The body is *for-itself* insofar as it is directed *towards* the world *from* the focal point (or point of view) of the subject.

It is important to note that the distinction between the body-object and an object that is in-itself may not be a fixed distinction. Merleau-Ponty discusses the example of a blind person learning to 'find one's way among things' with a walking stick.[92] 'Once the stick has become a familiar instrument', Merleau-Ponty explains, 'the world of feelable things recedes and now begins, not at the outer skin of the hand, but at the end of the stick'.[93] At this point, 'the stick is no longer an object perceived by the blind man, but an instrument *with* which he perceives'.[94] The walking stick, at this point, becomes 'a bodily auxiliary'.[95]

Perception, for those with sight and for those relying on a walking stick, is the 'organic relationship between subject and world',[96] and the body-object is what enables the relationship. A walking stick may initially be an object that has a 'mechanical relationship' between other objects. As such, the distinction between the body-object and an object (that is *in-itself*) would map onto the point where the skin of the hand meets the end of the walking

[90] M Merleau-Ponty, *Phenomenology of Perception* (n 69) 84–9.
[91] M Merleau-Ponty, *Phenomenology of Perception* (n 69) 115.
[92] M Merleau-Ponty, *Phenomenology of Perception* (n 69) 175.
[93] M Merleau-Ponty, *Phenomenology of Perception* (n 69) 176.
[94] M Merleau-Ponty, *Phenomenology of Perception* (n 69) 176.
[95] M Merleau-Ponty, *Phenomenology of Perception* (n 69) 176.
[96] M Merleau-Ponty, *Phenomenology of Perception* (n 69) 178.

stick. Once the walking stick is *unified with the body in response to the task of perception*, the walking stick becomes a 'bodily auxiliary' and the distinction between body-object and object will map onto the point where the end of the walking stick meets the world of objects.

I suggest that this reasoning can also apply in the other direction. A body-object may no longer be part of the subject's response to a task. To start with a trivial example, the hair that grows on the top of my head is part of the way I try to present myself, it is *for-me*. The hair that grows on the back of my neck frustrates the way I try to present myself, and I eagerly alienate it when I have my hair cut. In less trivial instances, the gametes of a couple attempting to conceive form an important part of a task, whereas the gametes of a couple seeking consequence-free sexual expression do not form part of a task (and may frustrate their task). What counts as the body is *self-ascribed* in this way: the subject determines what counts as the body *for-itself*. Whether an item of bodily material is *for-itself* or *in-itself* will depend on whether it is with the body in response to the task.

Most separated bodily material is likely discarded as mere-objects. Yet, if the body is understood in terms of responsiveness to tasks, the physical separation of bodily material may not change the functional relationship between the bodily material and the subject. It may remain directed towards a current or possible task, and thereby be functionally unified with the body in its responsiveness to the task of coping with, existing in, and engaging with the world. Bodily material may, like the walking stick, become a 'bodily auxiliary'. Here we can also identify a second-order aspect of our being. If the body or bodily material that is *for-itself* has a first-order value as the way in which a component of a person's subjectivity manifests itself in the external and physical world through a 'task', the process of self-ascription has a second-order value as the way in which we determine our tasks, determine the value attributed to tasks and conscript the body into a task.

e. The Body For-Itself and Self-Ownership

Since our 'subjectivity is physical', we cannot divide the world up into subjective experiences and objective bodies. Nor can we divide the world of objects into the body and other things. Items of bodily material (separate things) may continue to be functionally equivalent to an item of integrated bodily material (as part of the body). In other words, an item of bodily material may continue to be *for-itself*.

Where the separated bodily material continues to be functionally unified with the body in its responsiveness to a task, it is possible to extend the principle of self-ownership to separated bodily material. This is because physical separation has not interrupted the connection between the subject and the body. A person ought to be able to exercise rights of use and control over the body or bodily material where the body or bodily material is engaged in a current or potential 'task'. Alternatively, where separated bodily material is not functionally unified with the body in its responsiveness to a task, the bodily material is an object *in-itself*, and there remains no particular connection between the attributes and characteristics of the person and the bodily material.

Note that Merleau-Ponty would resist the description of the body as something that is 'owned'. As Taylor Carman explains:

as embodied perceivers, we do not typically understand ourselves as pure egos standing in a merely external relation to our bodies, for example by 'having' or 'owning' them, instead the body is itself already the concrete agent of all our perceptual acts.[97]

According to this view, *being* and *owning* are fundamentally distinct categories. Either something stands in external relation to our bodies (owning) or something is already engaged in a task or act (being). As S Kay Toombs suggests, the 'ownedness of the body is not a relation of possession but rather one of existing'.[98]

This, I would suggest, does not prevent us from applying the short-hand expression of 'self-ownership' to describe our relationship with our bodies and items of bodily material.[99] As Cohen explains:

The term 'self' in the name of the thesis of self-ownership has a purely reflexive significance. It signifies that what one owns and what is owned are one and the same, namely, the whole person.[100]

It is therefore not a contradiction to say that 'I' have 'full and exclusive rights of control and use' over 'my body'. Just as it is not a contradiction to describe someone as 'self-governing', 'self-legislative' or 'autonomous'. We can formulate these concepts without forcing a separation between the governor and the governed, the legislator and the legislated, or the *autos* (the 'self') and the *nomos* (the 'law'). Similarly, the principle of self-ownership does not rely on the separation or alienation of the 'owned-thing' from the 'owner'. We will nonetheless return to Merleau-Ponty's unease with the 'ownership' of the body when we compare property rights with the right to bodily integrity (specifically at pages 134–6).

[97] T Carman, 'The Body in Husserl and Merleau-Ponty' (n 70) 208.
[98] SK Toombs, 'What Does it Mean to Be Somebody' (n 78) 84.
[99] N Naffine, 'The Structure of Self-Ownership' (n 8) 194.
[100] GA Cohen, *Self-Ownership, Freedom and Equality* (n 7) 69.

It is therefore possible to identify a sound basis for why a progenitor ought to retain rights of use and control over his or her bodily material. Recall how the Hegelian 'dual nexus' is a sound justificatory basis as it is able to identify a set of instances where the connection between, or integration of, a person's body and his or her subjectivity is sustained. I have argued that where the bodily material retains a functional unity with the body, the integration of a person's body and that person's subjectivity is sustained. In such instances, where the bodily material is unified with the body in response to a task (bodily material that is *for-itself*) entitlements in bodily material ought to arise as an extension of a person's (reflexive) self-ownership.

f. The Body For-Others

Our 'physical subjectivity' has a further dimension. Since the subject is engaged in a physical-material world, when we interact with another subject, we encounter 'a fellow flesh-and-blood creature with whom we share a common' physical-material world.[101] Merleau-Ponty's account of other people 'rests on the recognition of the bodily medium of social perception, a medium common to myself and others'.[102] Our own bodies, as part of the physical-material world, become expressive of our subjectivity. As Toombs suggests, 'certain unique ways of moving, gesturing, speaking, laughing, and so forth, express *my* bodily being'.[103] Our bodies are the 'manifestations of behavior'.[104]

According to this view, the body is ambiguous insofar as the body is capable of having one or more meanings or value; the body is both an object *for-itself* whilst also an object *for-others*.[105] My perception and interaction with other subjects is caught within this ambiguity. The other is an object *for-others* to the extent that my social perception and interaction is mediated by their body-object. The other is not a mere object since 'my perception of the other' does not 'reduce' him or her 'to the status of an object'.[106] Others

[101] T Carman, *Merleau-Ponty* (n 70) 83.

[102] T Carman, *Merleau-Ponty* (n 70) 141: 'Merleau-Ponty's alternative account rests on a recognition of the bodily medium of social perception, a medium common to myself and others, *and which always already constitutes us as a community prior to our application of concepts such as mind and consciousness, which abstract from the bodily character of the person they describe*'. [emphasis added].

[103] SK Toombs, 'What Does it Mean to Be Somebody?' (n 78) 82.

[104] M Merleau-Ponty, *Phenomenology of Perception* (n 69) 352; S Priest, *Merleau-Ponty* (n 73) 189.

[105] M Merleau-Ponty, *Phenomenology of Perception* (n 69), 194–5; SK Toombs, 'What Does it Mean to Be Somebody?' (n 78) 82.

[106] M Merleau-Ponty *Phenomenology of Perception*, C Smith (tr) (Routledge, London 1962) 352; S Priest, *Merleau-Ponty* (n 72) 189.

subjects are therefore 'neither visible objects nor invisible subjects'.[107] When people die, their bodies lose a dimension. That is, the body is no longer the way in which they experience, and engage with, the world. A dead person has no subjectivity that engages with the world. However, the body remains an object that continues to be a medium of social perception. For all others, the body of a deceased person remains the 'manifestation' of the person. It is no longer *for-itself*, but it remains *for-others*.

If we understand our existence or 'being' as the sum of intentional ties or connections with the physical world, then included in this bundle of connections are the relationships that we form with other physical subjects. The body of another person therefore forms part of our existence or 'being'. Our relationship with another physical subject is something that has (first-order) value. Upon the death of that person, the body of the other remains as our connection with the person. Although we know that the deceased is no longer a subject, the death of the deceased does not reduce the body to the status of a mere object. There remains, therefore, a basis for explaining why the family or longstanding friends of a deceased person ought to be able to determine how the body of the deceased person is treated. The relationship between a parent and child, or between spouses, or between siblings, and so on, is a relationship mediated by the body. The posthumous body remains as the medium of the relationship. As the interim inquiry into the practice of organ retention at the Bristol Royal Infirmary found:

> For the parents of a recently deceased child, human material, certainly substantial specimens such as organs and parts of organs and even smaller parts are still thought of as an integral part of the child's body and thus, are still the child.[108]

The interest that family and longstanding friends have in the posthumous body of another may not be an indefeasible interest. There may be instances where it is outweighed by competing interests. Yet, their interest is more than just a 'relatively weak' and non 'person-affecting' interest.[109]

g. First-Person and Third-Person Perspectives

This philosophical turn, shifting from forging connections between inner subjects and external objects to the suggestion that our subjectivity is a

[107] T Carman, *Merleau-Ponty* (n 70) 150.

[108] Bristol Royal Infirmary Inquiry, 'The Report of the Public Inquiry into Children's Heart Surgery at the Bristol Royal Infirmary 1984–1995: Learning from Bristol' (Cm 5207, 2001) Annex C: 'Interim Report: Removal and Retention of Human Material' [33].

[109] J Harris, 'Law and Regulation of Retained Organs' (n 18) 548.

physical subjectivity, forms a foundational contrast for subsequent chapters. Allow me therefore to summarize the philosophical contention. The central claim here is that *the body is constitutive of the subject* insofar as a person, and the duties we owe the person, cannot be understood without reference to his or her body. This claim is in opposition to the position that the *body is contingent to the subject.* If the body is contingent to the person, the person's identity and the basis of our duties toward them can be understood without reference to the person's body.

There are three further propositions that I suggest follow from the contention that the body is constitutive of the subject. The first is that *what counts as* the body-subject ought to be understood in terms of the responsiveness of the body (or bodily material) towards an existing or possible 'task'. According to this view, separated bodily material may from part of the body-subject where it is functionally unified with the body in terms of a responsiveness to a task. If it is not functionally unified, the bodily material is *in-itself* (or *parte extra parte*). This is the first ambiguity in the body and bodily material, that the body is capable of more than one meaning insofar as it be *for-itself* or *in-itself.* In comparison, if the body is contingent to the subject, then the body and bodily material are (in all instances) mere instrumental objects or material resources that the subject, or others, may use to achieve certain ends.

If the body is the site, or location, of our subjectivity, the second further proposition is that the body is the medium of social experience. This is the second ambiguity in the body. A person's body is ambiguous insofar as it is both their own (*for-itself*) and exists for other people (*for-others*) where, upon the death of a person, the body of the deceased person may continue to be the medium of social experience. In comparison, if the body is contingent to the person, upon the death of that person his or her body remains as a mere object or surplus material resource.

The third further proposition is that our subjectivity (that which grounds moral duties) is un-abstracted. A person's rationally determined rules and standards, that person's welfare, utility, or preferences, his or her project, plans, goals, and interests (or whatever attributes, characteristics, or features of a person that solicit duties of respect) are tasks that are interwoven with the physical environment. The physical environment, prior to anything else, is a person's physical body. It is not possible, therefore, to abstract the basis of a person's value (their subjectivity) from the body and physical environment. The alternative proposition is to separate a person's subjectivity from the body, the result of which is to view the value of the body as a mere means to achieving some abstracted state of autonomy, utility, preference satisfaction, or 'human flourishing'.

Some may find these contentions difficult to accept. The problem, as Carman notes, is that 'from a third-person point of view, the structures and capacities of the body are mere contingent, arbitrary facts about the kinds of creatures we happen to be.'[110] From a third-person perspective, separated bodily material and walking sticks are merely instrumental objects, and dead bodies are merely masses of bodily material. Yet, as Carman also explains, the structures and capacities of the body are facts that 'can never manifest themselves *as* contingent and arbitrary *for us*, from our point of view, for they *are* our perspective on the world'.[111] Since the body is our point of view on the world, *on which we can have no point of view*, the structures and capacities of our bodies present themselves as necessary and constitutive.

This contrast between the first-person and third-person point of view nonetheless underlies and explains the disagreement over the use and legal status of bodily material. I suggest that the philosophical contrast can help explain the disagreement over whether progenitors (or their family members) ought to be able to control the use of bodily material (at the expense of the entitlements of healthcare and research institutions), whether the sale of bodily material denigrates the body, and whether the application of property law is the most appropriate legal structure to govern the use of bodily material. Hence, the two views of the body developed here will be relevant for a series of subsequent arguments. The first argument concerns how we address the tension between the entitlements that arise on a pre-social basis and the competing claims of use and control that are justified on a social basis. We can now turn to consider this tension.

III. Self-Ownership and Competing Social Needs

Even if we are able to establish that an item of bodily material engages a person's subjectivity, the value identified through connecting a person's subjectivity with their bodily material is not the only value that requires consideration. Since entitlements in bodily material can also be justified with reference to the valuable state of affairs that follow from the exercise of entitlements, we must also consider the social value of bodily material.

110 T Carman, *Merleau-Ponty* (n 70) 81.

111 T Carman, *Merleau-Ponty* (n 70) 82–3.

a. The Social Value of Bodily Material

In a number of important instances, there is a tension between competing interests in bodily material. The organ retention controversy in the United Kingdom is one such instance. According to the first-person view developed above, the body and bodily material of a deceased child may continue to be *for-others*. This can explain why the parents of a recently deceased child have an interest in determining how, and by whom, the body of the deceased child is treated. Even if we agree with this contention, we may nonetheless disagree over the strength of this interest relative to competing interests.

In the same year as the interim inquiry into the practice of organ retention at Bristol Royal Infirmary found that 'for the parents of a recently deceased child' bodily material was 'still thought of as an integral part of the child's body and thus, are still the child',[112] the Chief Medical Officer in his annual report stated that:

> There have been many occasions in the past where the study of tissue after death had led to discoveries in medical science which have resulted in the saving of lives and the relief of suffering.[113]

Given the social value in using bodily material as part of therapeutic, diagnostic, scientific, and educational enterprises, a tension arises between the social goods that can arise from the use of bodily material and the close connection between a progenitor (or a family member) and bodily material. In the context of the organ retention controversies, a pre-social analysis identifies the importance of the relationship between parents and newly deceased infants to the parents and would suggest that they retain control over the bodies and body parts of their children. A social analysis identifies the social value of using bodily material for diagnostic, education, and research purposes and would suggest that healthcare institutions ought to obtain the right to possess and use the body parts of deceased persons.

This tension arises in a range of contexts. It is difficult to doubt the moral value of providing bodily material to sustain or improve the life of another. Especially since organs, gametes, and tissue are in short supply. For instance, the Nuffield Council of Bioethics has observed that:

> three people die every day while waiting for an organ transplant; many fertility clinics are not able to meet requests for treatment involving donor eggs or sperm;

[112] Bristol Royal Infirmary Inquiry (n 108) [33].

[113] Department of Health, *Annual Report of the Chief Medical Officer* (2001) 6; J Herring, *Medical Law and Ethics* (4th edn, Oxford University Press, Oxford 2011) 413.

and research organisations cite difficulties in accessing bodily material as a key factor limiting research progress.[114]

The value of physicians and medical researchers being able to use material in a clinical context for diagnostic gain is equally incontestable. The potential to isolate the genetic sequences that are connected to cancerous cells, the ability to determine causes of death, and the ability to further identify the characteristics of diseases, conditions, or environmental factors that cause illness, suffering, and death, are all important—and life-saving—features of the provision of healthcare.

Hence, healthcare and research institutions having entitlements in separated bodily material may be justified on the basis of the therapeutic treatment that can be provided to patients, the knowledge obtained from diagnostic use of bodily material, and the developments that follow from using bodily material in medical research. To transfer an organ, to diagnose a health condition, or to identify the genetic features of cancerous tissue, requires the legal power to possess, manage, transfer, and use the bodily material. In order for these institutions to use the bodily material for therapeutic, diagnostic, and research purposes they need to obtain a large bundle of entitlements in the material. The tension between competing values arises because entitlements that are retained by progenitors of the bodily material, or their family members, limit the bundle of entitlements that healthcare institutions can exercise in bodily material.

For example, the requirement of consent or authorization for posthumous organ donation limits the availability of organs and bodily material for use in often life-saving treatment or valuable medical research. Which is why Aaron Spital and James S Taylor argue that the removal of cadaveric organs for transplantation ought to be a routine practice that bypasses the need for consent or authorization, on the basis that 'when one can save an endangered person at little or no risk to oneself, rescue is morally obligatory'.[115] In a similar vein, Harris argues that 'if we can save or prolong the lives of living people and can only do so at the expense of the sensibilities of others, it seems clear to me that we should'.[116] Any 'individualistic considerations', according to Harris and Antonia Cronin, 'must give way to utility in the interests of social and political morality'.[117] Harris, Cronin, Spital, and Taylor are right

[114] Nuffield Council on Bioethics, *Human Bodies: Donation for Medicine and Research* (October 2011) 3.

[115] A Spital and JS Taylor, 'Routine Recovery of Cadaveric Organs for Transplantation: Consistent, Fair, and Life-Saving' (2007) 2 *Clin J Am Soc Nephrol* 300, 303.

[116] J Harris, *Wonderwoman and Superman: The Ethics of Human Biotechnology* (Oxford University Press, Oxford 1992) 100.

[117] A Cronin and J Harris, 'Authorisation, altruism and compulsion in the organ donation debate' (2010) 36 *J Med Ethics* 627.

to identify the significant moral value in saving and prolonging the lives of others through provision and transplantation of bodily material.

Cecile Fabre offers a similar approach. Her argument draws upon a 'principle of sufficiency'[118] which requires that those who 'have less than what they need for a minimally flourishing life have a claim to surplus resources'.[119] By extension:

> if the materially needy have a right against the well-off that the latter transfer some of their income to them, then the medically needy have a right against the healthy that the latter transfer some of their body parts to them.[120]

Hence, when the principle of sufficiency is applied to bodily material, the contention arises that 'the needs of the sick should, in principle at least, be met by the confiscation of organs from the dead and even from the living'[121] provided the confiscation does not jeopardize the donor's prospects of a flourishing life.

b. The Body as a Material Resource

These views all share a common calibration between competing values. Avoiding the unnecessary loss of life and human suffering has greater moral magnitude than respecting the consent of the progenitor, the wishes of the deceased, or views of the family members, since the pre-social interests are either severely limited or do not arise in the first place. Harris makes explicit the way he calibrates the competing values. He describes the interests of progenitors or family members as 'the sensibilities of others'[122] and suggests that to prioritize these interests 'involves the equivalent of sacrificing people's lives so that others will simply *feel* better or not feel so bad, and this seems nothing short of outrageous'.[123] This conclusion follows from Harris' view that the dead have 'no will, no preferences, wants nor desires, the dead cannot be autonomous and so cannot have their autonomy violated'.[124] For Fabre

[118] C Fabre, *Whose Body Is It Anyway? Justice and the Integrity of the Person* (Oxford University Press, Oxford 2006) 30–8; C Fabre, 'Reply to Wilkinson' (2008) 14 *Res Publica* 137, 140.

[119] TM Wilkinson, 'The Confiscation and Sale of Organs' (2007) 13 *Res Publica* 327, 328: 'Fabre claims that justice includes a principle of sufficiency and she argues from this principle to the confiscation of body parts. Sufficiency says that people who have less than what they need for a minimally flourishing life have a claim right to surplus resources. A minimally flourishing life is an autonomous life, so it is more than a life where basic needs, such as for food and shelter, are met.'

[120] C Fabre, 'Reply to Wilkinson' (n 118) 140.

[121] TM Wilkinson, 'The Confiscation and Sale of Organs' (n 119) 327.

[122] J Harris, *Wonderwoman and Superman* (n 116) 100.

[123] J Harris, *Wonderwoman and Superman* (n 116) 100.

[124] J Harris, 'Law and Regulation of Retained Organs' (n 18) 531.

herself, the living body has value insofar as it enables 'human flourishing'.[125] It is then unclear for Fabre why our interest in bodily integrity is valuable in its own right. Given the premise that bodily integrity does not have a standalone value, Fabre's analysis raises the question as to why the right to bodily integrity should outweigh the needs of potential recipients for organs.[126]

For these commentators, there are no conflicting moral duties. On one hand, we are caught by duties of rescue and the duties that follow from the principle of sufficiency. On the other hand, the dead body has no relevant moral qualities and therefore does not attract duties of respect, and our respect of a person's bodily integrity is limited to where the right to bodily integrity is acting as a means of enabling a minimal standard of 'human flourishing'. These commentators are able to resolve the tension by reducing the value of the body and bodily material to a 'material resource'.[127] It follows that the body and bodily material, as a material resource, can be routinely retained without consent and redistributed in order to address healthcare needs in a way that is akin to a system of coercive taxation.

c. Abstract and Physical Flourishing

This view that the body is generally equivalent to any other material resource stems from an abstracted conception of subjectivity. If we view the body as contingently associated with the person or subject, then the body only has value as a means of enabling autonomous action, experiencing states of welfare, or generally providing for the conditions for human flourishing. Bodily material that is surplus to the minimal conditions for human flourishing is therefore a 'surplus resource'. Equally, if we view the body as contingently associated with the person or subject, then the body of a deceased person (who is no longer autonomous, experiencing, or flourishing) is also a 'surplus resource'. In essence, the body as a material and surplus resource ought to be put to *best use;* blood, marrow, and organs ought to be transplanted, tissue ought to be retained and researched, gametes ought to be made available for reproductive therapy and research, and so on. In other words, if we accept the external body is distinct from the 'inner' subject, then Fabre is entirely

[125] C Fabre, *Whose Body Is It Anyway?* (n 118) 110.

[126] TM Wilkinson, 'The Confiscation and Sale of Organs' (n 119) 333: 'To hold that bodily integrity is valuable in its own right, she thinks, produces an impasse. Someone who thought it so would not be persuaded by her arguments that justice can require confiscation, and she does not know what to say to such a person, except that, if both bodily integrity and a minimally flourishing life are independently valuable, it is not clear why bodily integrity should outweigh the needs of potential recipients for organs.'

[127] TM Wilkinson, 'The Confiscation and Sale of Organs' (n 119) 332.

correct to question why the bodily integrity of one person should outweigh the medical needs of others. Moreover, Harris is entirely correct to question why the sensitivities of some people should outweigh the medical needs of others. Such views, however, rest upon the abstraction of the subject from the body.

Yet, it is not clear why we should view a person's subjectivity as something that is detached and abstract. 'Personhood', according to Fabre, can be understood in terms of overlapping chains of psychological continuity.[128] This psychological criterion may be able to identify the qualities or features of our personal identity (qualitative personal identity) and address questions concerning personal identity over time (numerical or diachronic identity).[129] However, even if we identify the qualities that we consider to be constitutive of personal identity, it does not necessarily follow that these qualities solicit a set of moral duties toward that person. There is an additional enquiry into the moral relevance of qualities or features that construct our personal identity. Hence, to address questions of value or moral duties, Fabre suggests that our basis for respect for persons is human flourishing. A 'minimally flourishing life' requires 'some opportunity to form, pursue, and potentially to revise a plan of life'.[130] To connect these ideas, we need to infer that a person's value or basis of respect is somehow an 'inner' or 'psychological' phenomena, which is concerned with the formulation of a 'plan of life', and is related to chains of psychological continuity.

In comparison, recall Hegel's insight that the 'purely abstract will' must develop a 'content of determinative ends'.[131] We pursue a plan of life, not in the abstract, but in the messy external, and physical world. The further insight that Merleau-Ponty adds is that 'having a body is, for a living creature, to be intervolved in a definite environment'.[132] There is no 'inner' person into which we can we retreat. The body, I have contended, is interwoven or integrated into a person's 'plan of life'; it is what enables the 'opportunity to form, to pursue, and potentially to revise a plan of life', and it is what enables the experience of plans, goals, and preferences. As a result, there is ultimately no logical space between our 'human flourishing' and our right to bodily integrity (or the principle of self-ownership). To the extent that human flourishing

[128] C Fabre, *Whose Body Is It Anyway?* (n 118) 12; D Parfit, *Reasons and Persons* (Clarendon Press, Oxford 1984) 207.

[129] See J McMillan, 'Identity, Self, and Dementia' in Hughes et al (eds) *Dementia: Mind, Meaning and the Person* (Oxford University Press, Oxford 2006).

[130] C Fabre, *Whose Body Is It Anyway*? (n 118) 32.

[131] GWF Hegel, *Philosophy of Right* (n 41) [34].

[132] M Merleau-Ponty, *Phenomenology of Perception* (n 69) 94.

represents the 'opportunity to form, pursue, and potentially to revise a plan of life', where the formation, pursuit, and revision of a plan of life is a series of phenomena that are integrated or interwoven with the body, the right to bodily integrity maps directly onto the value of human flourishing. Since there is no logical space between our human flourishing and our bodily integrity (or self-ownership), our bodily integrity cannot be a means to enable the ends of a flourishing life. Instead, one is representative of the other.

The reason why bodily integrity is valuable in its own right is because the right to bodily integrity protects both first-order and second-order aspects of our being. It protects a person's ability to pursue a plan or preference which entails the use and control of their body (the first-order value). It also protects the ability of a person to formulate plans and preferences which entails the ability to recruit the body as *for-itself* (or alienate an item of bodily material as an object *in-itself*). That is, to be self-ascriptive (the second-order aspect of our being). Hence, some bodily material may nonetheless represent a surplus resource. Where the bodily material is not engaged in a current or possible task, the bodily material is an object *in-itself* and has no particular connection to the subject. As a surplus resource it ought to be put to *best use*. To that end, Fabre's principle of sufficiency is applicable to bodily material that is *in-itself*.

The distinction between the body that is *in-itself* and the body that is *for-itself* is a 'self-ascribed distinction'. It is up to each person to determine his or her own projects or 'plan of life' and conscript his or her body into the 'plan'. It is therefore a feature of my (reflexive) self-ownership that I can determine whether my hair or gametes are part of my subjectivity or the external world of objects, just as it is a feature of my (reflexive) self-ownership that I can allow or refuse a surgical incision into my body, or allow or refuse a sexual advance. It is this ability of a person being the primary arbiter of what is done to his or her own body that has value in its own right. Since the 'opportunity to pursue a plan of life' or to 'engage in projects' is only possible through the body, respect for someone's 'human flourishing' or subjectivity must entail respect for that person's rights of use and control of his or her body.

Moreover, the body is both *for-itself* and *for-others*. Given that our 'plans of life' or our 'human flourishing' include forming relationships with other people, these bonds or intentional ties are mediated through our bodies. These relationships, like our will, autonomy, or preferences, cannot exist in the abstract. They are lived relationships. The relationship of control over the body of the deceased touches upon a core aspect of our being or flourishing, that is, how we express ourselves in our relationship with other embodied beings. Although the way in which we express ourselves as a participant in a relationship will vary, the value of being able to express ourselves as a parent, as a child, as sibling, or as a spouse, has value in its own right.

d. Conflicting Duties

To the extent that we are forced to account for two conflicting duties, between duties of respect for a person's self-ownership and duties of rescue, there is no straightforward way to resolve the tension. First and foremost, I have contended here that (following the first person view of the body) *there is* a tension because there are two values that require consideration: the body (and in some instances, bodily material) has value as our medium of being in the world or as the medium of social experience, and the body (and bodily material) has value as a material resource that can be used to sustain, prolong, and improve the lives of others. Viewing the body and bodily material from this (first-person) perspective does not resolve the tension between conflicting duties; there remains difficult decisions about how to best balance competing and incommensurable values. The first-person view does nonetheless have a series of implications that help explain the contours of recent legal developments, and perhaps our intuition, over these disputes.

First, bodily material can be an object *in-itself*. When cancerous cells are removed from the spleen,[133] tissue is removed from the prostate,[134] when gametes are transferred as a unconditional gift[135] the bodily material is not unified with the body in response to a task and only has instrumental value. Such bodily material represents 'surplus resources' and can be treated as any other material resource. An implication of this view is that where bodily material is removed, there is no pre-social justification for why the progenitor ought to retain any rights of control or use. In these instances, the principle of sufficiency is applicable. That said, since a dimension of a person's rights of control and use over their body is that the subject may determine what bodily material is *for-itself* and what bodily material is *in-itself*, the requirement of informed consent remains critically important to respecting the principle of self-ownership.

Second, where the bodily material retains a functional unity with the body in terms of a responsiveness to current or potential tasks, the bodily material remains within the sphere of a person's self-ownership. Where a person's bodily material retains a functional unity with the body, a person's rights of exclusive use and control over his or her body are extended over the bodily

[133] *Moore v Regents of the University of California* (n 29).
[134] *Washington University v Catalona* 437 F Supp 2d 985 (Dist Ct Missouri, 2006).
[135] *CC v AW* 2005 ABQB 290.

material. In the same way that competing social interests are seldom considered to be sufficient enough to outweigh a person's self-ownership or right to bodily integrity, a person's entitlements in his or her body or bodily material that is *for-itself* ought to also be near-immune to competing social interests. Since we are caught by two duties (duties of respect for self-ownership and duties of rescue), we inevitability confront a balancing exercise between the relative weight of these duties. My suggestion is that our duty of respect for a person's self-ownership over his or her own body, and bodily material that retains a functional unity, is a duty that ought not to be displaced by simply pointing to resource shortages. The balancing exercise that we are engaged in here is disanalogous to coercive taxation or the redistribution of surplus goods.[136]

Tensions arise even if we were to limit ourselves to those interests that arise in a pre-social analysis. For instance, in *Evans v Amicus Healthcare*[137] and in *Davies v Davies*,[138] both male and female progenitors had a pre-social interest in the embryos since their gametes retained a functional unity with their bodies. The courts resolved this conflict by prioritizing one task (non-procreation) over a competing task (procreation). Further ethical analysis, in terms of which 'task' or component of the progentitors' subjectivity ought to have priority, is ultimately required to resolve this tension. The contention here has merely been that tension arises because the embryos in these cases are the site or location of the subjectivity of the progenitors.

The third implication is that where the body or bodily material of a deceased person continues to be the medium of social experience, the family and friends of the deceased have a pre-social interest in determining how the body of the deceased is treated. Note that, according to this view, the deceased retains no posthumous interest in the body. It is in this context that the tension between pre-social and social interest is difficult to resolve since two significant and incommensurate values are in play: an individual's interest in exercising some control over the body of a loved-one and the social interest in furthering the provision of healthcare through the availability of bodily material.

The tension stems from the difference between the third-person and first-person points of view. For example, from the point of view of the physicians and researchers at Alder Hay, Liverpool Children's, Bristol Royal

[136] TM Wilkinson, 'The Confiscation and Sale of Organs' (n 119) 332: '[T]here is then no obvious inconsistency with endorsing coercive taxation of material resources while rejecting coerced transfer of body parts or personal services.'

[137] *Evans v Amicus Healthcare Ltd and others* [2004] EWCA Civ 727.

[138] *Davies v Davies* 842 SW 2d 588 (1992).

Infirmary, Southampton University, and John Radcliffe Hospitals,[139] the bodies of infants were 'contingent' and 'arbitrary' facts about deceased children. As a result, the bodily material represented 'surplus resources' that could be used in the performance of socially valuable enterprises. For the parents of the deceased children, the bodily material of the children was 'an integral part of the child's body and thus, are still the child'.[140] Since the subjectivity or identity of a person can never express or extend itself to us without the body of that person, the body of another can never appear, or 'manifest' itself, as contingent and arbitrary to the person. Just as our bodies are (a first-person) point of view on the world that we can have no (third-person) point of view on, the bodies of others are our (first person) point of view of a subject, that we can have no (third person) point of view on.

The same tension arises in the context of the donation of bodily material. Those who donate bodily material to enable organ, blood, and marrow transplants, assisted reproductive therapies, and biomedical research are performing at least a morally praiseworthy act. For bodily material that represents a mere material resource, then the principle of sufficiency and duty of rescue ought to apply, and the donation of *surplus* bodily material (bodily material that is *in-itself*) is more than praiseworthy. Rather, the provision of bodily material that is *in-itself* would be morally obligatory. However, I have argued here that our self-ascribed bodies are more than surplus material resources, and our control over our bodies is not a means to further value. The implication is that those who donate bodily material that is *for-itself* or *for-others*, are prioritizing another's physical subjectivity over their own. In these instances, they are performing a morally praiseworthy, but not obligatory, act.

IV. Chapter Summary

This chapter has sought to identify when, if at all, a person ought to be able to retain the ability to use and control his or her bodily material. Although we all may agree that a person has rights of use and control over that person's own body, it is less clear how and to what extent the physical separation of

[139] See: Bristol Royal Infirmary Inquiry, 'The Report of the Public Inquiry into Children's Heart Surgery' (n 108); K Liddell and A Hall, 'Beyond Bristol and Alder Hey: The Future of Regulation of Human Tissue' (2005) 13 *Med L Rev* 170; JK Mason and GT Laurie, 'Consent or Property? Dealing with Body Parts in the Shadow of Bristol and Alder Hey' (n 26); *Re: Organ Retention Group Litigation* [2005] QB 506.

[140] Bristol Royal Infirmary Inquiry, 'The Report of the Public Inquiry into Children's Heart Surgery' (n 108) Annex C: 'Interim Report: Removal and Retention of Human Material' [33].

bodily material from the body, or the death of the person, changes the status of the bodily material.

The current common law approach is to identify entitlements of possession, use, and control where there has been the application of work or skill (the Lockean basis under *Doodeward v Spence*) or where the bodily material has been extracted by the progenitors for their own benefit (the Hegelian basis under *Yearworth v North Bristol NHS*). Although these are reasonably sound legal tests for the recognition of entitlements in bodily material, I have argued that we need to consider a more fundamental question in order to fully grasp why a person retains the ability to own bodily material. That is, we need to consider whether the relationship between a person and his or her body (or bodily material) is an arbitrary and contingent relationship, or a necessary and constitutive relationship. This question will re-emerge in subsequent chapters. I have sketched in this chapter a view of the body that is necessary to, and constitutive of, the person by suggesting that our subjectivity is a physical subjectivity. It follows from this view that the body is the medium of our engagement with the world (the body is *for-itself*) as well as the medium of social experience (the body is *for-others*).

To explain the contrast between these two conceptions of the relationship between a person and his or her body I adopted Carman's distinction between the first-person and third-person perspective. Since we can only ever have a first-person perspective of our body, the body will always present itself to us as necessary and constitutive. The same is true for the body of those we form relationships with. Hence, as will be a reoccurring theme, disagreement over the use and legal status of bodily material stems from the distinction between the first-person and third-person perspectives.

We encountered the first instance of this clash of viewpoints when we identified the tension between the pre-social interests and social interests in the use and control of bodily material. From the third-person perspective, the body and bodily material are mere contingent facts about the person and therefore form mere material resources. As material resources, bodily material ought to be redistributed in accordance with the principle of sufficiency or the duty of rescue. Although there are some instances where the bodily material is a surplus material resource (bodily material that is *in-itself*), I have suggested that we ought to resist viewing the body as something detached from our free-floating subjectivity. Our own bodies (and the bodies of loved-ones) are not mere contingent facts about us, nor are they a means to a further ends of 'human flourishing'. Following Merleau-Ponty, 'having a body is … to identify oneself with certain projects and be continually committed to them'.[141] Our bodies are the

[141] M Merleau-Ponty, *Phenomenology of Perception* (n 69) 94.

way in which we engage with the world and others. In other words, the body is the way in which we flourish.

To return, finally, to the six cancer patients in *Yearworth v North Bristol NHS*, the Court of Appeal's explanation for why the patients obtained the right to use their bodily material aligns with the Hegelian justification of the ownership of things. It is because the semen acquired an attribute which can only be explained 'in terms of the working will',[142] and because the semen enables an intended state of affairs that would otherwise not exist without the acquired attribute, that the patients had the right to use the semen. The semen, I suggest, is akin to Merleau-Ponty's example of a blind man's walking stick;[143] the samples of semen were not merely objects physically external to the patients but a way in which they engage in the world that is functionally equivalent to the body. It is because of this functional unity with the body that the patients' self-ownership can be extended over the separated bodily material to justify their rights of use and control. Moreover, the patients' relationship with the semen is a relationship that is particular to, and necessarily associated with, them. As we will consider in subsequent chapters, it is because the content of the patients' ownership relationship with their semen is necessarily associated with them that property law is ill-suited to protect the type of ownership relationship that arose in *Yearworth v North Bristol NHS*.

[142] J Waldron, *The Right to Private Property* (n 27) 369.
[143] M Merleau-Ponty, *Phenomenology of Perception* (n 73) 176.

3

The Commodification of Bodily Material

Dr David Golde diagnosed Mr John Moore as having hairy-cell leukaemia at the UCLA Medical Centre. He recommended that Moore have his spleen removed. Before the operation, Golde 'formed the intent and made arrangements to obtain portions of [Moore's] spleen following its removal and take them into a separate research unit'.[1] Moore travelled from Seattle to the UCLA Medical Centre several times over the subsequent seven years to provide 'additional samples of blood, blood serum, skin, bone marrow aspirate, and sperm' whilst under the impression that 'such visits were necessary and required for his health and well-being' and that such procedures could only be performed under Golde's direction.[2]

In fact, Golde was conducting research on Moore's cells and planned to benefit financially from having exclusive access to his bodily material. Golde developed a patentable cell line from Moore's T-lymphocytes. 'Genetic Institute' paid Golde $300,000 (in addition to $75,000 of shares, salary, and fringe benefits) in exchange for exclusive access to the 'materials and research performed on the cell line and product derived from it'.[3] The potential value of the cell line was assessed at over $3 billion.

John Moore was wronged. His legal right to make autonomous medical decisions was infringed by Golde. Without his informed consent, Golde used Moore as a means to his own commercial and professional ends. However, according to the majority of the Supreme Court of California, Moore did not have a legal right to a share of the profits from the use of his bodily material. In essence, Moore's rights of self-ownership were limited to rights of use and control over his body, and could not be extended to the right to profit from the use of his bodily material. We can now investigate this limit on the rights of self-ownership as it applies to bodily material.

[1] *Moore v Regents of the University of California* 793 P 2d 478 (Cal 1990) 126.
[2] *Moore v Regents of the University of California* (n 1) 126.
[3] *Moore v Regents of the University of California* (n 1) 127–8.

The purpose of this chapter is to assess whether the progenitor of bodily material ought to be able to profit from the transfer or relinquishment of bodily material to another individual or institution. Following the distinction between pre-social and social justifications, I will first assess whether the right to profit can be justified with reference to the characteristics or attributes of the person (a pre-social justification). I will then consider the circumstances in which the right to profit can be justified with reference to the state of affairs that the exercise of the right will produce (a social justification).

In terms of a pre-social analysis, I will argue here that the considerations which provide content to the right to profit are of a different order of consideration than the characteristics and attributes of the person. It follows that attempts to justify the right to profit on a pre-social basis are unsuccessful because an explanatory gap inevitably emerges between the considerations that enable the exercise of the right to profit and the characteristics or attributes of the person. It is therefore not possible to justify the right to sell or profit from bodily material on a pre-social basis.

It may, however, be possible to justify the right to profit in a social analysis. The provision of a financial benefit or reward may motivate more people to transfer or forgo their bodily material, and the resulting increase in the available bodily material may enable a greater number of therapeutic treatments and further enable scientific research. I will argue here that the availability of financial gain will motivate more people to transfer bodily material where there is a value equivalence between the value of the bodily material and the value of the financial benefit. Because of the ambiguity in the body, the value of an item of bodily material to the progenitor (or family member) may either be equivalent to, or non-equivalent to, a monetary value.

I will then suggest that where bodily material has, for the progenitor or family member, a value that is non-equivalent with a monetary gain, this value is *denigrated* if it is treated as being a value that is equivalent with the value of the monetary benefit. The denigration of something that is value non-equivalent, or valuable in its own right, can indicate a higher-order problem: that the way in which we attribute value to things may be tainted or vitiated. In terms of the sale of bodily material, financial pressures may taint the way in which a person is self-ascriptive. I will contend that our decisions as to what aspects of our bodies form part of 'us' are decisions that ought to be insulated from financial pressure.

I. A Pre-Social Right to Profit

In the previous chapter we assessed whether a person ought to be able to control and use his or her own bodily material. In this chapter, we engage in a separate assessment. Here we will consider whether a person ought to be able to profit from the provision of bodily material to another person. I have separated these assessments for the reasons outlined in chapter one: since control rights and the right to profit represent 'two importantly different aspects of ownership which must be considered separately and justified according to contrasting considerations',[4] the task of explaining why an individual or institution ought to be able to exercise an entitlement in bodily material will vary depending on the entitlement in question.

a. The Right to Profit and the Synchronic Connection

Recall John Christman's distinction between 'control rights' and 'income rights'. Control rights represent the various relationships of possession, use, and control of an object or resource. They concern, therefore, a relationship 'between an owner's set of preferences—her will—and the general state of the object owned'.[5] Income rights, in comparison, represent the ability of a person to financially benefit from the transfer or relinquishment of entitlements in an object or resource. The right to profit therefore concerns the transfer of entitlements in exchange for value. What differentiates the right to profit from control rights is the additional element that features when an entitlement-holder exercises her right to profit. That is, the financial benefit that is provided in exchange for the transfer of the entitlement.

The content of the financial benefit that is provided in exchange for the object or resource is determined by a set of additional factors, such as the abundance or scarcity of the particular resource, the demand for the resource, and the transaction costs of acquiring the resource. These factors are dependent on a structure that is external to the entitlement-holder, the recipient of the entitlement, and the rules protecting the entitlements. That is, a set of economic and social factors. The key point here is that the value or benefit that is derived from exercising the right to profit is determined by factors beyond the object or resource, the transferor, and transferee.

[4] J Christman, 'Self-Ownership, Equality and the Structure of Property Rights' (1991) 19 *Polit Theory* 28.

[5] J Christman, 'Distributive Justice and the Complex Structure of Ownership' (1994) 23 *Philos Public Aff* 225, 232.

This connection between the value or *benefit* that is derived from exercising the right to profit and the *distribution of resources* in society is twofold. The first connection is that the relative abundance or scarcity of a resource (as well as the demand for the resource and the presence or absence of transaction costs in acquiring the resource) determines what a possessor of the resource may obtain in exchange for the resource. This is a 'synchronic connection' (a connection occurring at a specific point in time) between the allocation of resources in society and the right to profit from a particular resource (the 'diachronic connection' is considered at pages 89–91).[6]

For example, if kidneys and blood were both tradable commodities, the profit received for the sale of a kidney would be higher than for the equivalent amount of blood because of the scarcity of the resource and the willingness of purchasers to pay for the kidney. In such cases, the financial benefit received is determined by the current distribution of resources. The rights-holder therefore receives a benefit that is the product of a range of factors external to the object and external to the entitlement-holder.[7]

Hence, the function of the right to profit is to enable the entitlement-holder to obtain a financial reward in exchange for entitlements in the object or resource. The financial reward is connected to a series of factors that are external to the rights-holder and the object or resource. The series of factors that give content to the right to income are social and economic considerations and, as I will demonstrate, these social considerations are logically distinct from the pre-social consideration of the characteristics and attributes of the person.

b. Self-Ownership and the Right to Profit

In the previous chapter we considered how the principle of self-ownership posits that 'each person enjoys, over himself and his powers, full and exclusive rights of use and control' provided 'that each other person also has just such a right'.[8] These rights, we can infer, extend into all spheres of life, including the economic sphere. It would then appear to follow that, since 'the autonomous individual is also very much an economic individual', the individual 'is free to contract for his labour in the market place'.[9]

[6] J Christman, 'Distributive Justice' (n 5) 228.
[7] J Christman, 'Distributive Justice' (n 5) 232.
[8] GA Cohen, *Self-Ownership, Freedom and Equality* (Cambridge University Press, Cambridge 1995) 213.
[9] B Bennett, *Health Law's Kaleidoscope: Health Law Rights in a Global Age* (Ashgate, Aldershot 2008) 95.

Since we are able to profit from the use of our bodies—contract as an economic individual in the market place—an intuitive analogy arises between using the body in the performance of a paid task and transferring bodily material in exchange for payment. The analogy may be then further affirmed when we consider the similarity of risks that arise between some medical procedures, such as a kidney transplant, and some paid jobs, such as deep-sea fishing.[10] The contention is concisely put by Julian Savulescu, who claims that:

> People have a right to make a decision to sell a body part. If we should be allowed to sell our labour, why not sell the means to that labour? If we should be allowed to risk damaging our body for pleasure (by smoking or skiing), why not for money which we will use to realise other goods in life?[11]

The similarity between the sale of a body part and paid labour is that the self-owner in both instances is exercising their self-ownership to make choices regarding his or her body and the body is used as a tool to perform a task for which the self-owner is financially rewarded. However, I will argue here that the right to profit cannot be justified in this way. That is, it cannot be justified as a right formulated with reference to the principle of self-ownership, or more broadly, it cannot be justified with sole reference to the characteristics and attributes of the person.

Let us start with the analogy. As self-owners we may undertake the health risks associated with smoking or skiing, we may choose to be photographed for publication, and we may exercise our labour to paint a portrait. Yet few of us obtain a financial reward from smoking, skiing, being photographed, or even painting a portrait. This is because there is an abundance of smokers, skiers, models, and artists. Our inability to profit from these choices regarding our body is not a limit on our rights of self-ownership. To say that I have the right to freely direct my body or to use skills or talents, or to say that I may agree with another that I will direct my body, skills, and talents in a particular way, does 'not entail that I thereby also have the right to benefit from the exchange of skills in any way available'.[12] The right to obtain a *financial benefit* from the use of the body does not naturally follow the attributes or characteristics of the person.

[10] JS Cameron and R Hoffenberg, 'The Ethics of Organ Transplantation Reconsidered: Paid Organ Donation and the Use of Executed Prisoners as Donors' (1999) 55 *Kidney Int* 727; C Fabre, *Whose Body Is It Anyway? Justice and the Integrity of the Person* (Oxford University Press, Oxford 2006) 131.

[11] J Savulescu, 'Is the Sale of Body Parts Wrong?' (2003) 29 *J Med Ethics* 138.

[12] J Christman, 'Self-Ownership, Equality and the Structure of Property Rights' (n 4) 32–3.

We cannot, therefore, presume that not affording the right to profit from the use of the body is inconsistent with the idea of self-ownership. In response to this presumption, Christman argues that 'preventing me from reaping these increased benefits [from the use of my talents] does not *ipso facto* prevent me from controlling my life'.[13] The mistaken presumption here is that the right to profit is a logical extension of the principle of self-ownership, or that the right to profit otherwise follows from the attributes or characteristics of the person. It is mistaken because it overlooks how the right to profit is a functionally distinct right. It is necessarily a social right since its *content* is determined by a series of social factors.

Those who try to justify the right to sell bodily material as flowing from the rights of self-ownership (or the natural attributes and characteristics of the person) obscure the distinction between considerations that are relevant to the *content* of the right to profit and the considerations relevant to the rights of control by focusing on the *process* of selling a body part. As Savulescu puts it, '[p]eople have a right to make a *decision* to sell a body part'.[14] By focusing on the preferences and choices with regards to the state of the object—the decision whether to sell the body part—the income right appears to fall into the ambit of a person's exclusive decision making authority over the person's own body.

Yet, decisions as to the general state of the object, such as whether body material is extracted, retained, or transferred for therapeutic or research purposes, are decisions that may be afforded to the progenitor with reference to his or her autonomy, his or her welfare, or preference satisfaction. The move from the 'decision to transfer a body part' to the 'decision to sell a body part' only involves the introduction of a financial reward in exchange for the transfer. To say that someone has the right to sell a body part, as something beyond affording them the right to transfer a body part, is to say that they have a right to the *outcome* of the transfer: the financial benefit. By focussing on the process of exercising an income right, proponents of a natural right to profit from the use or provision of bodily material can only be justifying the right to transfer bodily material. A pre-social justification cannot be extended to explain why the transferor has the right to financially benefit from the transfer.

We have been able to explain the principle of self-ownership by connecting a person's autonomy, utility, or preference satisfaction (or other possible characteristics or attributes of the person) with rights of control and use of

[13] J Christman, 'Self-Ownership, Equality and the Structure of Property Rights' (n 4) 34.
[14] J Savulescu, 'Is the Sale of Body Parts Wrong?' (n 11) 138 [emphasis added].

the body. However, in order to justify the right to profit from the use or relinquishment of the body or bodily material, we must be concerned with considerations beyond the (pre-social) attributes and characteristics of the person. That is, social considerations as to the current and desired distribution of bodily material.

c. The Dual Nexus and the Right to Profit

Given this preliminary conclusion—that we are unable to ground the right to profit in the principle of self-ownership—let us assess whether this argument applies to more nuanced, but still pre-social, accounts of ownership. Let us take Jeremy Waldron's use of the Hegelian basis of entitlements in *The Right to Private Property*[15] as an example of a successful basis for justifying the recognition of control rights (as discussed at pages 50–6) and consider whether it is able to connect the right to profit with the attributes and characteristics of the person.

Waldron offers an argument for the recognition of property rights which assumes that the Hegelian-owner of a thing also owns its *value*.[16] Waldron explains the basis of property rights in terms of a 'nexus of dual relations between the will and object, object and will'.[17] The second of the two relationships between the object and will requires that the object is capable of introducing a state of affairs that is willed, or intended, by the individual. Hence, if an object acquires new attributes through the intentional acts of an individual, and the individual does so intending a state of affairs where the object can be alienated in exchange for an income, then the right to profit may become a legitimate part of the Hegelian account of ownership.

For instance, Matthew may bake three-dozen croissants with the intention of selling them. The sugar, butter, and flour have acquired new attributes through the intentional acts of Matthew and now the croissants can introduce a new state of affairs. In this case, the state of being exchanged for income. Similarly, Lucy may extract ova with the intention of selling it. By her body, Lucy alone generated the ova. Lucy, through her intentional actions, has extracted the ova. The external state of the ova now enables Lucy to profit from the provision of the ova. In both instances there is an arguable dual relationship between 'will' and 'object' that includes the state of affairs where the object can be exchanged for a financial benefit.

[15] J Waldron, *The Right to Private Property* (Clarendon Press, Oxford 1988) 343–423.
[16] JW Harris, *Property and Justice* (Oxford University Press, Oxford 1996) 245.
[17] J Waldron, *The Right to Private Property* (n 15) 370.

To explain why the above Hegelian explanation cannot succeed in justifying the right to profit, allow me to return to a preliminary principle. For any pre-social justification there needs to be a connection between the attributes and characteristics of the person that demand our respect (the justification) and the relationship with the object that is being asserted (the entitlement). Moreover, the connection between the justification and the entitlement needs to be a proportional connection. Otherwise, the explanation or justification for why an individual is an entitlement-holder may be under-extensive; the justification will be too weak to be able to support the full set of entitlements in the object or resource that are being exercised.

According to the Hegelian basis, respect for the person includes respect for the connection between the person and an object created by a 'dual relationship between will and object'. For example, if Matthew had an object in his possession that invokes poignant memories, Joanna's respect for Matthew would involve a respect for Matthew's sentimental relationship with that object. In terms of entitlements, Matthew would have a right to possess the sentimental object (and Joanna would be under a correlative duty with regards to Matthew's possession). Joanna's respect for Matthew requires the recognition of an entitlement that is necessary for the manifestation of Matthew's personhood in the external world.

'But this respect', Alan Carter argues, 'should not be confused with property rights' in the fullest sense.[18] We may use property rights to require the respect for someone's rights of possession, use, or management, but it does not follow that the need to 'extend oneself' into the external world gives rise to a full bundle of ownership entitlements that include the right to income. The entitlements that we ought to recognize are determined by the necessary conditions for the expression of personhood, and therefore the content of the ownership-bundle is circumscribed by the *reasons* for recognizing the entitlement.

On one hand, Joanna respects Matthew's exclusive rights of possession over the sentimental object because Matthew's possession of the object is necessary for Matthew's personhood to be expressed through the sentimental object. In this case, the right of possession is able to protect the 'dual relationship between will and object' by ensuring that you are able to retain access to poignant memories. Matthew would not, on the other hand, have the right to transfer the sentimental object. Were he to have such a right, there would be a dislocation between the entitlement and the reason for the entitlement.

[18] A Carter, *The Philosophical Foundations of Property Rights* (Harvester Wheatsheaf, London 1989) 137.

What this example attempts to illustrate is that the entitlement that an individual has in an object is limited to the entitlements that are necessary in order for the Hegelian individual to 'express their personhood' or sustain the 'dual nexus'. Otherwise, without the connection between the entitlement and reason for the entitlement, justification is under-extensive and there is no reason for the imposition of a duty on others to respect the entitlement.[19]

d. The Missing Step in the Justification

The reason why it is not possible to formulate a right to profit in a pre-social analysis is because there will *always* be a disconnect between the right to profit and the pre-social basis for entitlements in things. Recall the functional distinction between control rights and income rights. Control rights concern the preferences and choices vis-à-vis the state of the thing or object. Income rights include a value or benefit that is constructed by a range of factors external to the ownership relationship, such as the demand for the resource, transaction costs, and the distribution of the resource in society. To justify the right to profit requires consideration of the external factors that determine the content of the entitlement, and the Hegelian basis of ownership only concerns factors that relate to the individual's will and the attributes of the object owned.

There is a gap, therefore, between the factors concerning the justified thing (the right to profit) and the factors used in the justification (the attributes and characteristics of the person as they relate to the object). Hence, when it comes to the right to profit it is difficult to see any connection between the expression or manifestation of personhood and the acquisition of a financial benefit. As James W Harris remarks (quoting Waldron), 'in what sense can cash or bank accounts constitute objects which can be seen, over time, to be "registering the effects of the willing"?'[20] It is difficult to identify an element of a person's will, personhood, or subjectivity that, in order to be expressed in an external object, requires the recognition of the ability to profit from the use or relinquishment of that external object. This is because the content of the right to profit (or the 'value' of the asset or resource) is determined by factors external to the entitlement holder. Again, a gap emerges between the factors in *what needs to be explained* and the factors that are available in any possible pre-social *explanation*.

The gap always emerges because of the distinction between the right to profit and pre-social reasons. As above, the content or value of an income

[19] A Carter, *The Philosophical Foundations of Property Rights* (n 18) 101.
[20] JW Harris, *Property and Justice* (n 16) 245.

right is dependent upon the distribution of resources in society. As a result, it appears contradictory to have a natural or pre-social right to an entitlement, which has value only insofar as the distribution of resources in society *gives* it value. The content or value of the right is 'distributionally determined',[21] yet natural rights are only concerned with the attributes of the person and are thus 'distributionally blind'.[22] Hence, there is 'a missing step in the justification—between the right to act freely and the entitlements to … benefit in a certain distribution of goods'.[23] The problem is that the right to profit cannot follow from the attributes and characteristics of the person, independent of social aims. Rather, social considerations, external to the attributes and characteristics of the person, are required to bridge the gap between the individual and their body, and income entitlements. Christman summarizes the contention:

The components of ownership that comprise of personal control and consumption of a resource must be justified according to importantly different considerations from those that justify the right to gain income from the resource … the source of the justification for income rights, however, will necessarily be principles that govern the pattern of distribution of goods in the economy, considerations which are not reducible to individual interests.[24]

People do not, therefore, have a right to make a decision to sell a body part that is analogous to decisions to risk damaging our body for pleasure. We may have the right to make a decision to transfer or relinquish a body part, but the right to profit from the transfer cannot be explained as a right that follows from our self-ownership or any other pre-social justification. Rather, the right to profit is necessarily a social right. In order to assess whether a person ought to be able to profit from the transfer or provision of bodily material to another, we must turn to a social analysis.

II. A Social Right to Profit

If the right to profit is necessarily a social right, let us consider whether the sale of bodily material can be justified with reference to the state of affairs that commercial exchanges in bodily material would introduce. Although there are settled

[21] J Christman, 'Distributive Justice and the Complex Structure of Ownership' (n 5) 231.

[22] JW Harris, *Property and Justice* (n 16) 246.

[23] J Christman, 'Can Ownership Be Justified by Natural Rights?' (1986) 15 *Philos Public Aff* 156, 173.

[24] J Christman, 'Distributive Justice and the Complex Structure of Ownership' (n 5) 230. See also J Christman 'Can Ownership Be Justified by Natural Rights?' (n 23) 174.

commercial practices in non-transplantable bodily material between healthcare or research institutions,[25] disagreements arise as to whether a monetary reward (beyond compensation of expenses and loss of earnings)[26] ought to be afforded to individuals in exchange for their bodily material. We will focus on this disagreement. In this section I will identify a possible social justification for the right to profit from the transfer or relinquishment of bodily material and identify circumstances in which the justification holds.

a. The Right to Profit and the Diachronic Connection

Recall that there are two connections between the value or benefit that is derived from the right to profit from an object or resource and the distribution of that resource in society. We have already encountered the first, 'synchronic', connection (the abundance or scarcity of a resource determines what a possessor of the resource may obtain in exchange for the resource).[27] The second connection is a 'diachronic connection'[28] (a connection over different points in time). This second connection suggests that the exercise of the right to profit that enables the object or resource to be lawfully exchanged, determines a new pattern of resource distribution.

The right to profit, or the absence thereof, effects the future distribution of resources by 'directing (or allowing) the distribution of goods to flow in some direction or other'.[29] The ability to acquire a good in exchange for a market-determined price has a significant effect on who has which resources, and hence, determines the future distribution of the resources in society. To return to the kidneys example, the permitted sale of kidneys would lead to a different (for better or for worse) allocation of kidneys in society since they could be exchanged, in a free or restricted market, as a commodity. The financial benefit may motivate the transfer of kidneys, and as a result, may better distribute the scarce resource.

Commentators rely on this positive diachronic connection to justify the right to exchange bodily material for financial reward. For instance, James S Taylor argues that:

> both experience and economics tell us that the additional number of organs that would be procured through the market would be greater than the resulting drop

[25] R Hardcastle, *Law and the Human Body: Property Rights, Ownership and Control* (Hart Publishing, Oxford 2009) 113; Hansard HC col 115 (28 June 2004) (Ms Winterton).

[26] Human Tissue Act 2004 (HTA 2004), s 32(7); Human Fertilisation and Embryology Authority, *Code of Practice* (8th edn, London 2013) [13].

[27] J Christman, 'Distributive Justice' (n 5) 228.

[28] J Christman, 'Distributive Justice (n 5) 228.

[29] J Christman, 'Self-Ownership, Equality and the Structure of Property Rights' (n 4) 32–3.

in the number of organs that would be procured by more traditional, non-market, methods.[30]

The suggestion is that the new pattern of distribution (the greater availability of organs) provides the normative basis for why a person ought to be able to profit from the transfer or relinquishment of bodily material.

As we have encountered in the previous chapter, according to the Nuffield Council of Bioethics, 'three people die every day while waiting for an organ transplant' and 'many fertility clinics are not able to meet requests for treatment involving donor eggs or sperm'.[31] To address these 'health needs' the Council recommends a series of 'interventions' that include 'financial incentives that leave the donor in a better financial position as a result of donating'.[32] Again, the normative basis for the ability of a progenitor to financially benefit from the transfer or relinquishment of bodily material is an allocative outcome.

b. The Diachronic Connection and Reasons for Action

The positive diachronic connection suggests that different structures of financial rewards or incentives will result in changes in social behaviour. In this context, we are concerned with the relationship between the introduction of a financial benefit and the motivations that inform the decision to relinquish or transfer bodily material. Such decisions by individuals are shaped by 'reasons for action'. These 'reasons for action' or motivations operate as 'the premises of an argument the conclusion of which is that there is reason for the agent to perform the action or that he ought to do it'.[33] 'Motivations' may include beliefs as well as facts. To use Joseph Raz's example, carrying an umbrella would be justified on basis that 'it is either the fact that it will rain or my belief that it will which would be cited as the reason'.[34]

For instance, a decision to donate the organs of a deceased loved-one may be motivated by lives that may be saved through an organ transplant.[35] The

[30] JS Taylor, *Stakes and Kidneys: Why Markets in Human Body Parts Are Morally Imperative* (Ashgate, Aldershot 2005) 3.

[31] Nuffield Council on Bioethics, *Human Bodies: Donation for Medicine and Research* (October 2011) 3.

[32] Nuffield Council on Bioethics, *Human Bodies: Donation for Medicine and Research* (n 31) 167–70.

[33] J Raz, *Practical Reason and Norms* (Oxford University Press, Oxford 1999) 15.

[34] J Raz, *Practical Reasons and Norms* (n 33) 17.

[35] M Sque, T Long, S Payne, and others, 'Why Relatives Do Not Donate Organs for Transplants: "Sacrifice" or "Gift of Life"?' (2008) 61 *J Adv Nurs* 134; L Siminoff, MB Mercer,

decision to leave the body of the deceased intact may be motivated by the belief that the deceased would not have wanted his or her organs donated.[36] Identifying such reasons for action would help us predict or anticipate changes in social behaviour that would follow from the introduction of a financial benefit in exchange for bodily material. Ultimately, we are unable to, from our armchairs, identify such reasons for actions. Such an inquiry is a sociological inquiry. We can nonetheless identify the circumstances under which there would be a positive diachronic connection between the right to profit from the transfer of bodily material and availability of bodily material.

I suggest in this section that the positive diachronic connection would hold to be true if:

(i) the provision of a financial benefit represents a reason in favour of transferring or relinquishing bodily material;
(ii) that is a reason for action sufficient, in the aggregate of cases, to outweigh the existing reasons in favour of retaining control or interring bodily material;
(iii) without also substantially undermining the existing reasons in favour of the donation of bodily material.

If such a connection holds then, given the reported shortages of bodily material, the greater availability of bodily material is an outcome that can (everything else being equal) justify the exercise of the right to profit.

c. Reasons for Action and Value Equivalence

According to the diachronic argument, the provision of a financial or monetary benefit will motivate the transfer or relinquishment of bodily material. This is because money has proven to be, in most instances, an effective motivator. Yet, consider how money provides a particular type of motivation or reason for action. As Michael Walzer explains, 'money is both the measure of equivalence and the means of exchange; these are the proper functions of money and (ideally) its only functions'.[37] It is because money functions as a 'means of exchange' that it has no value in its own right. It only has

G Graham, and others, 'The Reasons Families Donate Organs for Transplantation' (2007) 62 *J Trauma* 969.

[36] M Sque, T Long, S Payne, and others, 'Why Relatives Do Not Donate Organs for Transplants' (n 35); L Siminoff, MB Mercer, G Graham, and others, 'The Reasons Families Donate Organs for Transplantation' (n 35).

[37] M Walzer, *Spheres of Justice: A Defence of Pluralism and Equality* (Basic Books, New York 1983) 104.

instrumental or substitutional value in that it enables us to obtain or achieve an innumerable set of other things. Moreover, it is because money functions as a 'measure of equivalence' that it enables us to calibrate the value of different things (with more accuracy than the direct comparison between two different things).

The provision of a financial benefit provides an *extrinsic* reason for action. When someone acts on an extrinsic reason for action, they are motivated by the 'underlying preference... for the reward associated with performing the task' or action.[38] The diachronic argument therefore relies upon extrinsic reasons for action, in the form of monetary benefits, motivating the transfer or relinquishment of bodily material. Moreover, when someone acts *rationally* on an extrinsic reason for action, they identify a 'value equivalence' between the reward associated with performing the task and the loss, costs, or burdens associated with performing the action. For instance, when Matthew decides to produce and sell his croissants to Mark, the amount Mark paid for the croissants ought to be value equivalent to the costs of producing the croissants. As David Bollier explains, a sale of a thing seeks to 'establish a "moral equilibrium" between parties to the transaction'.[39]

As we can anticipate with regards to Matthew's croissants, the gain, reward, or benefits from performing the action need not be equal to, or be the same as, the loss, costs, or burdens associated with the action. If they were, then the action would be value- or reasons-neutral. In fact, we would expect that a rational actor would act on extrinsic reasons where the benefit is, for them, greater than the burden. The 'value equivalence' that is required for an action to be motivated by an extrinsic reason is that the benefit or reward that is obtained from the act is from the 'sphere of value' or has the same 'mode' of valuation,[40] as the loss or burden suffered from the act. Croissants, in this way, have a value equivalence with the price paid for the croissants.

Where extrinsic reasons are unable to motivate an action is where the benefit or reward associated with the action and the reasons that oppose the action belong to different spheres of valuation. Although we all value money, and although we all value friendship, we value these things in a different way, using a different mode of valuation. The value of money and the value of friendship belong to different spheres of valuation; 'it is the nature

[38] BS Frey and R Jegen, 'Motivation Crowding Theory' (2001) 15 *J Econ Surv* 589, 591; following EL Deci, 'Effects of Externally Mediated Rewards on Intrinsic Motivation' (1971) 18 *J Pers Soc Psychol* 105.

[39] D Bollier, *Silent Theft: The Private Plunder of Our Common Wealth* (Routledge, New York 2002) 40.

[40] E Anderson, *Value in Ethics and Economics* (Harvard University Press, Cambridge 1993) 114.

of friendship that it cannot be bought'.[41] We therefore have difficulty contemplating questions that force an equivalence between these two values, such as: how much money we would pay for an additional friend, or for how much money would we forgo a friendship? Although it may be possible to rationally answer these questions (if given an inflated exchange rate) such answers nonetheless 'do violence to the way we think [such things] should be properly valued'.[42]

Hence, some things that we value have an equivalence with other things. It is because of this equivalence that we can forgo some these things in favour of others. When we act on an *extrinsic* reason for action (ie when we perform a task because of the rewards associated with the task) we identify a value equivalence between what we forgo in the performance of the task and what we gain from the performance of the task. It follows that where a reward is associated with a task, the reward will motivate the performance of the task if the reward has value equivalence with, at least some of, the existing reasons not to perform the task.

In contrast, other things that we value have value non-equivalence. They have value that is not substitutional with another thing of value. If the reasons for performing a task stem from a value that has value non-equivalence, then reasons for performing the task may be *intrinsic* reasons. That is, performing a task on the basis that the task has value in its own right (independent of what else can be obtained from the action).[43] We can anticipate that, in a number of instances, people will view their relationship with their bodies, items of bodily material, or the bodies of loved-ones, as having a value that is not substitutional with another thing of value. Our subjectivity (however constituted) has value its own right, and to the extent that our subjectivity is integrated or interwoven with our bodies, and to the extent that our relationships with others are relationships with embodied beings, it would follow that our bodies and the bodies of others may have a value that is not substitutional with another thing of value. Note that it is difficult, in comparison, to view the body or bodily material as having value in its own right if we adopt the third-person perspective. If our subjectivity is abstracted, free-floating, and on a level of personal experience that is separate from our sub-personal physiological systems and capacities, then the body and bodily material may only have substitutional value as one of the possible means of achieving these 'higher states' of autonomy, welfare, or preference satisfaction.

[41] G Cohen, 'The Price of Everything, The Value of Nothing: Reframing the Commodification Debate' (2003) 117 *Harv L Rev* 689, 704.

[42] G Cohen, 'The Price of Everything' (n 41) 697.

[43] BS Frey and R Jegen, 'Motivation Crowding Theory' (n 38) 591.

d. Intrinsic and Extrinsic Reasons for Action

What makes predicting the positive diachronic connection between the availability of a monetary reward and the resulting availability of bodily material difficult is the interaction between extrinsic and intrinsic reasons for action. If a decision to relinquish or transfer bodily material was only governed by extrinsic reasons for action and inaction, then the greater the monetary reward for the bodily material, the greater the extrinsic reasons there are to relinquish or transfer the bodily material, and the greater the likelihood that each individual would relinquish or transfer bodily material.

For example, we may view our posthumous body, or the body of a deceased loved-one, as 'no longer needed'; as having no particular value in its own right.[44] We may also view our regenerative bodily material, such as our blood, as having no intrinsic value, and cite the 'time and inconvenience' associated with donating blood when explaining why we do not regularly donate to a blood bank.[45] In these circumstances, the positive diachronic connection may hold to be true. This is because, in such circumstances, there are no intrinsic reasons for retaining control of the entire body of the deceased nor any intrinsic reasons for retaining control over bodily material. The ability to financially benefit will generate reasons to relinquish or transfer bodily material that are capable of defeating any existing reasons for *in*action.

What complicates the diachronic connection is that we may view the value of our body, our bodily material, or the body of a deceased loved-one as having value in its own right. In some instances, extrinsic and intrinsic reasons for actions may be concurrent reasons for action. We may view the provision of bodily material to others as a valuable act independent of what we gain from the transfer. Although such intrinsic reasons for action may belong to a different sphere of value to financial (extrinsic) motivations, the different types of motivations may nonetheless concur. For example, a provider of gametes may be motivated by the desire to 'help infertile couples' and assist others in their reproductive therapy by providing some of the necessary bodily material.[46] This assistance may be viewed as having value in its own

[44] M Sque, T Long, S Payne, and others, 'Why Relatives Do Not Donate Organs for Transplants' (n 35); L Siminoff, MB Mercer, G Graham, and others, 'The Reasons Families Donate Organs for Transplantation' (n 35).

[45] TW Gillespie and CH Hillyer, 'Blood Donors and Factors Impacting the Blood Donation Decision (2002) 16 *Transfus Med Rev* 115; GB Schreiber, KS Sclumpf, SA Glynn, and others, 'Convenience, the Bane of Our Existence, and Other Barriers to Donating' (2006) 46 *Transfusion* 545; M Harrington, MR Sweeney, K Bailie, and others, 'What Would Encourage Blood Donation in Ireland?'(2007) 92 *Vox Sanguinis* 361–7.

[46] KR Daniels, R Curson, and GM Lewis, 'Semen Donor Recruitment: A Study of Donors in Two Clinics' (1996) 11 *Hum Reprod* 746.

right. The provider of the gametes may find concurrent motivation to provide the gametes 'for financial reasons'.[47] In such instances, the provision of gametes is motivated by concurrent intrinsic and extrinsic motivations, and again, the diachronic connection may also hold to be true: potential providers of bodily material will be motivated by financial reward.

However, concurrent reasons for action may result in motivational over-crowding. The problem arises where the intrinsic value of performing an act is tainted by an external intervention. In a well-known study, Richard Titmuss has indicated that this may be the case with regards to blood donation in his comparative study between the United States and the United Kingdom.[48] Recent studies have also identified a crowding-out effect in payment for blood.[49] The implication is that, even where there are concurrent intrinsic and extrinsic reasons in favour of an action, the 'external intervention' of the extrinsic reasons for action may displace the intrinsic reasons for action and reduce the overall availability of a resource. The external intervention can be explained as an instance of 'impaired self-determination', where 'an agent perceives an external intervention as reducing their self-determination' since the agent is 'forced to behave in a specific way by outside intervention'.[50] Alternatively, the crowding out may be the result of 'impaired self-esteem' where 'outside intervention carries the notion that the actor's [intrinsic] motivation is not acknowledged' or unable to be exercised.[51] The crowding-out of intrinsic reasons for action may provide a negative diachronic connection, where the availability of a financial benefit decreases the supply of bodily material.

Importantly, extrinsic and intrinsic reasons for action may *conflict*. We may find that the intrinsic value of performing an action is at the expense of an extrinsic benefit or reward, or *vice versa*. For instance, the reasons why relatives do not donate the organs of a deceased family member most often include the perceived duty of guardianship to protect intrusions into the body, the fear that donation would disfigure the body, the belief that the deceased would not want to have his or her organs donated, the desire to be present for the 'observable' death of the patient, and the belief that the

[47] R Almeling, 'Why Do You Want to Be a Donor?: Gender and the Production of Altruism in Egg and Sperm Donation' (2006) 25 *New GenetSoc* 143.

[48] RM Titmuss, *The Gift Relationship: From Human Blood to Social Policy* (Allen & Unwin, London 1970).

[49] C Mellström and M Johannesson, 'Crowding Out in Blood Donation: Was Titmuss Right?' (2008) 6 *J Eur Econ Assoc* 845.

[50] BS Frey and R Jegen, 'Motivation Crowding Theory' (n 38) 594.

[51] BS Frey and R Jegen, 'Motivation Crowding Theory' (n 38) 594.

patient or the family have been through enough pain and discomfort.[52] These are intrinsic reasons to retain control over the body of the deceased. For the families of deceased persons, the act of performing the perceived duty of guardianship, or being present for the observable death of the person, has value in itself. Were the family members offered a financial reward for consenting to the removal of organs, they would be caught between different types of reasons, originating from different spheres of value.

How this conflict of reasons is resolved depends on the relative strength of reasons. For some, the 'belief that the deceased would not want to have their organs donated' may provide a conclusive reason[53] to retain control. Their reasons for retaining control of the entire body would then be immune to financial incentives. For others, the extent of their financial need may provide sufficient reason to forgo control of the bodily material despite the intrinsic value of the body. The key point here is whether payments will motivate the further provision of bodily material depends, in part, on whether the value of the bodily material has 'value equivalence' with money. Where there are intrinsic reasons for action (or inaction), such reasons may limit or undermine the positive diachronic connection. Intrinsic reasons for *in*action may conflict with extrinsic reasons for action, or even when the intrinsic and extrinsic reasons for action are concurrent reasons, the extrinsic motivation of financial benefit may crowd-out the intrinsic motivations.

e. The Ambiguous Body and Value Equivalence

In the previous chapter I described the body, and bodily material, as being ambiguous. Since the body is both a *subject* and *an object*, bodily material is ambiguous insofar as it may be an object *for-itself* or an object *in-itself*. Whether an item of bodily material is *for-itself* and part of the subject, or whether it outside the sphere of subjectivity and part of the world of things, will depend on whether the bodily material is engaged in a current or potential task (where a 'task' represents the integration of a component of subjectivity with the physical world). The body is, in this way, 'self-ascribed' by the subject. Hence, as Margaret Radin identifies:

> we have things that are ours, but they are objects, or at least we can detach them from ourselves without destroying the notion of personhood, and when we do that we can trade them in markets.[54]

[52] M Sque, T Long, S Payne, and others, 'Why Relatives Do Not Donate Organs for Transplants' (n 35); L Siminoff, MB Mercer, G Graham, and others, 'The Reasons Families Donate Organs for Transplantation' (n 35).

[53] J Raz, *Practical Reasons and Norms* (n 33) 27.

[54] MJ Radin, 'What, If Anything, Is Wrong with Baby Selling?' (1995) 26 *PAC LJ* 135, 143.

We may, in some circumstances, consider our blood, tissue, or gametes to be outside the sphere of our subjectivity since it does not assist us in the tasks of coping with, existing in, and engaging with, the world. The bodily material may have no particular or intrinsic value to us. It may have value insofar as we can obtain something else in exchange for it. In which case it has a substitutional value or a value that has value-equivalence, and decisions with regards to bodily material that is *in-itself* will be responsive to extrinsic reasons for action, such as a financial reward or benefit.

However, in other circumstances, we may view our bodily material as forming part of the response to a task. In these circumstances, the bodily material may have value in its own right, since the bodily material is interwoven with a component of our subjectivity. Recall that this value is a first-order aspect of our being; it is the way in which we achieve or satisfy our goals and preferences. In addition, it is because the body and bodily material is ambiguous that the exercise of self-ascription—the process of determining what constitutes the body-subject or what items of bodily material continue to be *for-itself*—also has value. This is a second-order aspect of our being; it is the value of being able to determine what is valuable. By way of analogy, we may value the relationship with a newborn child, and consider the relationship to be inalienable and even our parental status as constitutive of our personhood.[55] The relationship with the newborn may be (first-order) valuable, and valuable in its own-right. Alternatively, we may not ascribe particular value to the relationship with a newborn child, we may consider the relationship with the child to be alienable, and consider making the child available for adoption or continue in the performance of a surrogacy agreement. What is, and what is not, valuable is matter of self-ascription. This exercise of self-ascription has a (second-order) value, and the value of being self-ascriptive is valuable in its own right.

To summarize the contention so far, this section has been concerned with the social basis for the exercise of the right to profit from the provision of bodily material. I have suggested that whether the claim that the availability of financial reward will increase the supply of bodily material depends on the interaction between extrinsic and intrinsic reasons for action. I have also suggested that there are two intrinsic values at stake. First, we may ascribe to our bodies or bodily material (that is *for-itself*) an intrinsic value since it is interwoven with our plans, goals, wishes, preferences (or any other component

[55] MJ Radin, 'What, If Anything, Is Wrong with Baby Selling?' (n 54) 143.

of our subjectivity). We may also ascribe to the bodies of others an intrinsic value since it is the medium through which we relate to others. Second, the process of determining our plans, goals, wishes, and preferences and how our bodies or bodily material enables and engages these components of our subjectivity, and the process of determining how we express our relationships with others, also has value in its own right.

There are a number of implications of this view. If we perceive an item of our body or bodily material to be outside the sphere of our subjectivity (to be an object *in-itself*), then the bodily material will not have any particular value to us. If it has any value, it will have a substitutional value; the value of the bodily material will be in what other thing of value the bodily material can obtain for us. Extrinsic reasons, such as financial reward, will provide a reason to relinquish or transfer the bodily material. In these instances, the diachronic justification for the right to profit from the relinquishment or transfer of bodily material may be applicable.

If we perceive an item of bodily material to be *for-itself* (or *for-others*), then it has a (first-order) value in its own right insofar as the bodily material is interwoven with a component of our subjectivity and that any such component of our subjectivity has value in its own right. The ability to profit from the provision of bodily material is *for-itself* (or *for-others*) and will create a conflict or a crowding between different 'spheres of value': the extrinsic value of the financial benefit and the intrinsic value of the bodily material. The conflict between the spheres of value, or the crowding of the spheres of value, may render the financial incentive to relinquish or transfer bodily material ineffective. In these instances, the diachronic justification would be unable to secure the social benefits that justify the exercise of the right to profit. Furthermore, as we shall now turn to consider, where there is a conflict between different spheres of value, the value of the bodily material may be 'denigrated'.

III. Value Denigration

There exists an uneasy division between what can and cannot be sold. As Gerald Cohen notes, (in the United States) 'sperm, art, pollution rights, and life insurance can be sold; votes, draft cards, and children cannot'.[56] Even if the sale of bodily material does increase the overall availability of bodily

[56] G Cohen, 'The Price of Everything' (n 41) 689.

material that can be put towards socially valuable ends, there may nonetheless be sound reasons for limiting what can be sold. In this section I will outline a general concern: that treating something that has value in its own right as somehow equivalent with a monetary payment denigrates the value of the thing. By placing this general concern in the context of the positive diachronic connection (between the right to profit and the availability of bodily material) we are then able to formulate a more particular concern: that the diachronic justification relies upon a forced value equivalence that taints the process through which we determine what is valuable to us or what is constitutive of us.

a. The General Concern: Value Denigration

When Immanuel Kant argues against the sale of a body part in *The Groundwork of Metaphysics of Morals* he claims:

> [I]n the kingdom of ends, everything either has a price or a dignity. If it has a price, something else can be put in its place as an equivalent; it is exalted above all price and so of not equivalent when it has dignity.
>
> What is related to general human inclinations and needs has a market price; that which, even without presupposing a need, conforms with a certain taste, that is, with a delight in the mere purposeless play of our mental powers, has a fancy price; but that which constitutes the condition under which alone something can be an end in itself has not merely a relative worth, that is, a price, but an inner worth, that is, dignity.[57]

We have already encountered the distinction that Kant is drawing here. An object that can be substituted for another object is an object that has 'an equivalent'. That is, a value equivalence. Since the value of the object is equivalent to the value of a substituted object, such objects 'have a price': the substitutional value of the object. An object that has no equivalent has 'dignity'. Since another object cannot be 'put in its place as an equivalent', an object with dignity cannot have a price. The value of an object with dignity is inherent to the object itself; it can be 'an end in itself'. It appears that, for Kant, the world is divided up into objects with a price and objects with an 'inner worth', and an object *either* has a price *or* an inner worth.

A number of commentators adopt this distinction in arguing against the commodification of bodily material. For instance, Stephen Munzer

[57] MJ Gregor (ed), I Kant, *The Groundwork of the Metaphysics of Morals* (Cambridge University Press, Cambridge 1996) 42–3.

expresses the concern that under some market arrangements, the language of the market:

> might distort the way in which people view themselves and others. They might tend to see persons as repositories of body parts with a market worth rather than entities with a Kantian dignity.[58]

Ruth Chadwick also expresses a similar concern that:

> one undesirable consequence of the selling of our bodies is that it contributes towards a society in which the bodies of persons are regarded as resources. The action of selling one's own body contributes to the prevailing ethos of everything being for sale, everything having a price.[59]

The underlying concern expressed here is that the sale of bodily material 'denigrates' the body as it 'treats items that belong in high spheres of valuation as if they were in the same sphere of valuation as money'.[60] The contention is that a person has an inherent or non-substitutional value, that this value ought not to be viewed as somehow equivalent to something of instrumental or substitutional value, and such an equivalence arises from the sale of bodily material. As Cohen explains in terms of the broader commodification debate, 'part of what makes an exchange improper is the transaction's expression of value equilibrium—that the things being exchanged are of equivalent value'.[61]

However, whilst we can say that a person's subjectivity has value in its own right, we cannot presume that an item of the body or bodily material also has an inherent, rather than substitutional, value. As Munzer himself recognizes, we cannot presume that 'what is true of the whole must be true of its parts'.[62] I have suggested that it is a matter of self-ascription whether part of the body or an item of separated bodily material is *for-itself* (and thereby retains the value in its own right) or is an object *in-itself* (and thereby has only substitutional or instrumental value). For a part of the body or an item of bodily material that is *in-itself*, there would be no value denigration were the bodily material to be sold. The substitutional or instrumental value of the bodily material that is *in-itself* has a *value equivalence* with the payment for the bodily material, since bodily material that is *in-itself* and a monetary payment belong to the same sphere of valuation.

[58] SE Munzer, 'Uneasy Case against Property Rights in Body Parts' (1994) 11 *Soc Phil & Pol'y* 259, 286.

[59] RF Chadwick, 'The Market for Body Parts: Kant and Duties to Oneself' (1989) 6 *J Appl Philos* 129, 137.

[60] G Cohen, 'The Price of Everything' (n 41) 696.

[61] G Cohen, 'The Price of Everything' (n 41) 689.

[62] SE Munzer, 'Uneasy Case' (n 58) 275.

Value denigration occurs when the value of an object that has value non-equivalence is treated as having an equivalent value. The value of this object is 'corrupted'[63] when 'the relevant goods cannot be aligned along a single metric without doing violence to our considered judgments about how these goods are best characterised',[64] but the relevant goods are nonetheless measured by a single metric. Consider two examples:

> If I buy a Nobel Prize, I corrupt the meaning of the Nobel Prize ... if I buy and sell children, I corrupt the meaning of parenthood.[65]

The value of a Nobel Prize is twofold: the 8 million Swedish Krona and the high esteem and societal recognition of an outstanding achievement. Leaving aside the Swedish Krona, the latter value of esteem and recognition has value in its own right. To buy a Nobel Prize, through bribing the relevant Assembly or Committee or from purchasing it outright from an previous award winner, would be to undercut or denigrate the (latter) value of the Prize. The value of the Nobel Prize is denigrated because *the way in which value is normally attributed* to the prize is tainted. Rather than the relevant Assembly or Committee identifying an outstanding achievement, the award is obtained through monetary payments.

Equally, if we value our relationship with a newborn child then the relationship will have, for us, value in its own right. If we were to then sell the child we would be *forcing* an equivalence between two values where one value (the value of the relationship) has no equivalence. In such a scenario, there is a forced value equivalence. The forced value equivalence is indicative of a higher-order problem. In this scenario, were we to forgo the relationship with a child in exchange for a monetary payment, we are forgoing our ability to ascribe to ourselves what relationships are (intrinsically) valuable to us.

In the same way, if we value our body or bodily material (that is *for-itself*) as having value in its own right, then the sale of any such bodily material is forcing an equivalence between two values where one value has no equivalence. This forced value equivalence is also indicative of a higher-order problem. In these circumstances, we are unable to be self-ascriptive; we are unable to determine what physical aspects of our existence are directed toward a

[63] Cohen, 'The Price of Everything' (n 41) 691.

[64] Cass R Sunstein, 'Incommensurability and Kinds of Valuation: Some Applications in Law' in R Chang (ed) *Incommensurability, Incomparability and Practical Reason* (Harvard University Press, Cambridge 1997) 234, 238.

[65] A Kimbrell, *The Human Body Shop: The Engineering and Marketing of Life* (Harpercollins, San Francisco 1993) 35.

certain existing or possible task (where a task represents the integration of a component of our subjectivity with the physical environment).

It is in this way that 'everything either has a price or a dignity'; either the bodily material has a value equivalence or it does not, and whether it has a value equivalence is a matter of self-ascription. In circumstances where we treat something that has value non-equivalence as having an equivalence, the value denigration may be indicative of a high-order problem. The contention here is that value denigration, or a forced value equivalence, may be a symptom of a tainted or vitiated process. In terms of the value denigration of bodily material, the process of self-ascription may be tainted. That is, the process through which we determine what connections or relationships with the world are constitutive of our subjectivity or personhood.

b. The Particular Concern: Financial Pressure and Value Equivalence

In the remainder of this section I will formulate a more particular concern with the sale of bodily material in light of the general concern for value denigration (outlined above). The particular concern follows from the threefold contention that: the positive diachronic connection relies upon the financial pressure, there are some processes that ought to be insulated from social or financial pressure, and that the process of being self-ascriptive is a process that ought to be insulated from social or financial pressure. This particular concern aims to provide a limit on the diachronic justification for the right to profit from bodily material.

The diachronic argument in favour of the right to profit from the relinquishment or transfer of bodily material either relies upon a value equivalence between the value of financial benefit and the value of the bodily material, or it relies upon the financial benefit providing an extrinsic reason for action that is capable of outweighing existing intrinsic reasons for inaction. In either case, if a financial benefit or reward is able to motivate or provide extrinsic reasons for action, then the monetary benefit would motivate progenitors or transferors of bodily material in a way that is proportionate to their financial need. In terms of the sale of organs, Simon Rippon argues:

> Those who want to sell their organs are not generally motivated by the promise of obtaining luxury goods such as holidays, recreational flying lessons or cases of fine wine, but by economic desperation.[66]

[66] S Rippon, 'Imposing Options on People in Poverty: The Harm of a Live Donor Organ Market' (2014) 40 *J Med Ethics* 145, 148.

Rather, it is 'because people in poverty often find themselves either indebted or in need of cash to meet their own basic needs and those of their families' Rippon predicts that they would respond to 'social or legal pressure to pay the bills by selling their organs'.[67] The assumption here is that only those who are living in financial need or deprivation are likely to be motivated to extract and transfer bodily material for monetary gain. If this assumption holds, then it follows that a market in bodily material *needs, or pre-supposes* that a category of people are living in financially constraining circumstances in order for the market to achieve the aim of increasing the availability of bodily material. As Paul Hughes explains, in response to Gerald Dworkin's proposal to limit the market in organs to those earning above 40% of the medium income, that:

Very few, if any, people above the lower 40% excluded from the market would ever wish to sell an organ (as Dworkin himself recognizes), so the effect of the 'paternalistic' measure... would be to eliminate not only the poor from the market, but the market itself. This means, of course, that *the poor are essential to the existence of a market in organs.*[68]

A person's financial depravity may, in some circumstances, be sufficient to motivate the sale of bodily material regardless of whether the bodily material has value in its own right. In less dire circumstances, making the option available to someone to alleviate financial pressure by the sale of bodily material may not, by itself, force the value equivalence and motivate that person to forgo their bodily material. However, a 'new "option" can... easily be transformed into a social or legal demand, and it can drastically change the attitudes that others adopt towards you'.[69] This explanation is premised upon Dworkin's idea of 'responsibility', which explains that:

Once I am aware that I have a choice, my failure to choose now counts against me. I now can be responsible, and be held responsible, for events that prior to the possibility of choosing were not attributable to me. And with the fact of responsibility comes the pressure (social and legal) to make 'responsible' choices.[70]

For example, consider a scenario where Matthew, a victim of domestic violence, is given the option of whether his partner and perpetrator of the violence, Mark, is prosecuted for the violence inflicted upon him. In such a

[67] S Rippon, 'Imposing Options' (n 66) 149.

[68] PM Hughes, 'Exploitation, Autonomy, and the Case for Organ Sales' (1998) 12 *Int J Appl Philos* 89, 94; responding to G Dworkin, 'Markets and Morals: The Case for Organ Sales' in G Dworkin (ed), *Morality, Harm, and the Law* (Westview, Boulder 1994) 155.

[69] S Rippon, 'Imposing Options' (n 66) 147.

[70] G Dworkin, 'Is More Choice Better Than Less?' (1982) 7 *Midwest Stud Philos* 47, 61.

scenario, the availability of a new option may place Matthew in a worse position than if the choice was not afforded to him. It is possible that there is sufficient reason not to afford Matthew the option. Although we cannot say, with any generality, that Matthew ought to permit the prosecution of his partner, we can nonetheless appreciate that the availability of the option to permit the prosecution can be the basis of social pressure. We can also appreciate that, in some circumstances, the social pressure to prevent the prosecution is pressure that is sufficient to vitiate the authenticity of the choice as to whether Mark is prosecuted. If there is social pressure on a victim of domestic violence, and the victim prevents the prosecution of the perpetrator, we can at least identify a significant risk that the process through which the victim arrived at their decision is a process that is tainted or vitiated by the social pressure.

The option to profit from the transfer or relinquishment of bodily material, like a victim's discretion as to the prosecution of their offender, is an option that can be transformed into a social or financial demand or can be the source of social or financial pressure. Moreover, the positive diachronic justification works on the basis that such demands and pressures increase the overall availability of bodily material. Hence, the positive diachronic justification reliance on financial benefits motivating decisions to relinquish or transfer bodily material is ultimately reliance on the application of financial pressure on those in constraining circumstances in order to increase the supply of bodily material.

c. The Sting in the Diachronic Tail

I therefore suggest that the positive diachronic justification of the right to profit has a sting in its tail. The justification works on the basis that a financial benefit or reward will motivate people to relinquish or transfer bodily material. However, such a financial benefit will motivate others in a way that is proportionate to their financial need. This raises the presumption that only those in financially constraining circumstances will be motivated to relinquish or transfer their bodily material *in response to* the availability of a financial benefit. So far, this by itself may be unobjectionable; we can identify a range of non-problematic instances where we have made decisions that we would not have otherwise made, but for our financial needs or constraints.

It is once we place the idea of value denigration in the context of decisions being made in response to financially constraining circumstances that concerns arise. Recall that a value is denigrated where a transaction 'treats items that belong in high spheres of valuation as if they were in the same sphere of valuation as money'.[71] The denigration of a first-order value may be

[71] G Cohen, 'The Price of Everything' (n 41) 696.

indicative of a second-order process being tainted. If we value something in its own right, and yet, in response to financial pressure, treat it as if it was 'in the same sphere of valuation as money', we denigrate the value of the thing. We can then infer that the financial pressure has tainted the process through which we attribute non-substitutional value to things.

If someone sold their newborn child, with whom a relationship has value in its own right, in order to alleviate financial pressure, something 'that belongs in high spheres of valuation' (the relationship with the child) is being treated as it belongs 'in the same sphere of valuation as money'. Moreover, the second-order value of being able to attribute value to a parental relationship is being tainted by the application of financial pressure. Although we rightfully permit a person to forgo a potential parental relationship (through adoption or surrogacy arrangements for instance), we may also rightfully attempt to insulate such decisions from being decisions that are made in response to financial pressure. Which is why a newborn is a 'type of good most likely to have unique value' that can adopted but not sold.[72] We accept that the value of parental relationships are self-ascribed, and therefore are alienable. We would also, I contend, view the decision whether to be a parent to be a process that ought to be insulated, as best as possible, from financial pressures.

In the same way, if someone relinquished control over the body of a deceased loved-one (which has value for that person in its own right) in response to an opportunity to alleviate financial pressure, then the first-order value of the relationship is denigrated through the forced value equivalence. Moreover, we can infer that the second-order value of being able to determine how they express their relationship has been tainted by the option of alleviating financial pressure.

If someone relinquishes or transfers bodily material that they would otherwise view as part of themselves, in order to alleviate financial pressure, the first-order value of their relationship with their body or bodily material is being denigrated through the forced value equivalence. Moreover, the second-order value of ascribing what forms part of them as an embodied subject, and what 'tasks' can be undertaken through their body, is being tainted. In this way we can understand 'why someone who sees absolutely no difficulty at all with transplantation and donation' can nonetheless 'sanction a sale' of bodily material.[73] We can accept that our bodies are self-ascribed,

[72] G Cohen, 'The Price of Everything' (n 41) 701.

[73] LR Kass, 'Organs for Sale? Propriety, Property, and the Price of Progress' (1992) 107 *Pub Int* 65, 77: '[I]t is difficult to understand why someone who sees absolutely no difficulty at all with transplantation and donation should have such trouble sanctioning sale.'

and that bodily material is therefore alienable, whilst also seeking to insulate the process of self-ascription from social and financial pressures.

The positive diachronic argument—that the increased supply of bodily material would be procured through the right to sell bodily material—presupposes that financial benefit will motivate the provision of bodily material. One implication of the availability of financial reward in exchange for bodily material is that the availability of the option translates into financial or social pressure on those living in financially constraining circumstances to relinquish, transfer, or forgo their bodily material. The general concern raised here is a concern for value denigration. If 'everything has a price' or a value equivalence, then there remains no logical space to value things in their own right. The particular concern raised here is that our decisions as to what aspects of our physical existence fall within our sphere of subjectivity (and our decisions as to what aspects of another person's physical existence remain the medium of social experience) are decisions that we ought to keep immune from financial pressure. If we allow financial pressure to enter into the equation, we are denigrating a core feature of our subjectivity. That is, we are allowing a calculus where a person's 'inner worth' (their autonomy, welfare, or preferences) is equated with the 'relative worth' of a monetary price.[74]

d. Limits on the Diachronic Argument

The diachronic argument may nonetheless be sound, in general: the availability of a financial benefit in exchange for bodily material may enable a state of affairs where more bodily material is available for a range of socially valuable uses. I have merely prescribed here two limits on the argument.

The first limit is that the ability to predict a positive diachronic connection is complicated by the interaction between intrinsic and extrinsic reasons for action. We cannot, as Taylor suggests, rely on 'experience and economics' in order to forge a positive connection between the right to profit and the greater availability of bodily material.[75] Rather, the positive diachronic connection may only arise where the financial benefit represents an extrinsic reason for action that does not 'crowd out' existing intrinsic reasons for action, and is able to defeat any conflicting intrinsic reasons for *in*action.

If the financial benefit is able to defeat any conflicting intrinsic reasons for *in*action, the second limit is that the social good of increasing the availability of bodily material may be off-set by the sting of value denigration. A value

[74] I Kant, *The Groundwork of the Metaphysics of Morals* (n 57) 42–3.

[75] JS Taylor, *Stakes and Kidneys* (n 30) 3.

is denigrated when two things that belong to different spheres of value are treated as if they belong to the same sphere of value. There may be a series of instances where retaining possession or control of bodily material represents a value that is derived from a different mode of valuation than the value of a financial benefit. Financial need may nonetheless force an equivalence between these different values. Given that the body is the site or location of our subjectivity, a forced value equivalence taints our ability to determine for ourselves what aspects of our physical existence are valuable in their own right and associate ourselves 'with certain projects and be continually committed to them'.[76]

The importance or relevance of these limits will vary depending on the bodily material that is being relinquished or transferred and the financial benefit that is being offered in exchange for the bodily material. The difficulty is that there are no fixed categories for which items of bodily material are *for-itself, for-others*, or *in-itself*. A person may view their hair, blood, gametes, regenerative tissue, and their posthumous body as an object *in-itself* (not engaged in a current or potential task). In such instances, the bodily material has a value equivalence with such things as financial benefit or reward. Where a person views their bodily material, as *for-itself* (engaged in a current or potential task) or views the posthumous body of another as *for-others* (the medium of social experience) the bodily material has value in its own right. The concern for value denigration arises here as the (first-order) value of the bodily material, and the (second-order) value in being self-ascriptive, exist in a different sphere of value from the financial reward, and that the financial benefit or reward taints our ability to be self-ascriptive.

The second variation is the financial benefit. In order to motivate or encourage the provision or relinquishment of bodily material, regulations may permit payments to the provider that fall short of a purchase price, such as compensation for expenses or loss of earnings.[77] The Nuffield Council on Bioethics, for instance, recommends an 'intervention ladder' with a series of six levels of public intervention, progressing from (1) 'information about the need for the donation of bodily material for others treatment or for medical research', (2) 'recognition of, and gratitude for, altruistic donation, through whatever methods are appropriate both to the form of donation and the donor concerned' through to (5) 'interventions offering associated benefits in kind to encourage those who would not otherwise have contemplated donating to

[76] M Merleau-Ponty, *Phenomenology of Perception* C Smith (tr), (Routledge, London 2003) 81, 94.

[77] HTA 2004, s 32(7); Human Fertilisation and Embryology Authority, *Code of Practice* (8th edn, London 2013) [13].

consider doing so' and (6) 'financial incentives that leave the donor in a better financial position as a result of donating'.[78]

Although the Council recommends financial incentives that leave the donor in a 'better financial position', the Council nonetheless seeks to avoid the item of bodily material being treated as equivalent to the financial benefit provided in exchange for the bodily material:

> where the intervention involves a direct payment of money or equivalent, it is an essential pre-requisite that the payment is understood, by all parties, in terms of reward to the *person* for their act of providing bodily material, rather than a purchase of material itself.[79]

This is a laudable aim. However, it may only be achieved if the amount of financial reward is directly tied to quantifiable losses incurred through the process of transaction (which would ultimately not 'leave the donor in a better financial position as a result of donating'). This is because, even if the parties to the transaction could understand the payment as a payment for the time, effort, and discomfort expended on the extraction and transfer of bodily material, this understanding between the two participants may not prevent the availability of the option to transfer bodily material for value from being understood by a wider category of persons as an option to use the body parts as a fungible commodity. It is the frequent exchanges of bodily material for a financial benefit that will create or reinforce a view that persons are 'repositories of body parts with a market value' or that 'bodies of persons are regarded as resources'.[80] If we are concerned about value denigration, then we ought to be concerned with the availability of the option to elevate financial pressure through the commodification of the body rather than isolated transactions between participants who view that monetary reward as reward 'to the *person* for their act'. One way to avoid this is to limit the monetary payment to compensation for losses and expenses incurred in the act of providing the material.

[78] Nuffield Council on Bioethics, *Human Bodies: Donation for Medicine and Research* (n 31) 167–70: 'Rung 1: information about the need for the donation of bodily material for others treatment or for medical research; Rung 2: recognition of, and gratitude for, altruistic donation, through whatever methods are appropriate both to the form of donation and the donor concerned; Rung 3: interventions to remove barriers and disincentives to donation experienced by those disposed to donate; Rung 4: interventions as an extra prompt or encouragement for those already disposed to donate for altruistic reasons; Rung 5: interventions offering associated benefits in kind to encourage those who would not otherwise have contemplated donating to consider doing so; Rung 6: financial incentives that leave the donor in a better financial position as a result of donating.'

[79] Nuffield Council on Bioethics, *Human Bodies: Donation for Medicine and Research* (n 31) 170.

[80] SE Munzer, 'Uneasy Case' (n 58) 286; RF Chadwick, 'The Market for Body Parts' (n 59) 137.

IV. Chapter Summary

There have been three main arguments advanced in this chapter. The first is that there is no pre-social basis for the right to profit from the provision of bodily material. My contention has been that the right to profit is constituted by a set of social and economic conditions that are beyond or external to the entitlement-holder. There is, therefore, a logical gap between the right to profit and the attributes and characteristics of the progenitor. In other words, considerations that contribute to the natural or pre-social rights of the person are of a different order of considerations that contribute to the right to profit or financially benefit from the provision or use of an object or resource.

We can appreciate why the majority of the Supreme Court of California in *Moore v Regents of the University of California* considered the consequences of allowing a patient to claim, through an action for conversion, a share of the profits from the unauthorized use of his spleen cells. The right to profit is a social right that is recognized by the law in order to secure social goals. In *Moore v Regents of the University of California*, the ability for the patient to claim a share of profits was seen as a restraint on the social goal of providing financial incentives to encourage socially valuable developments in biotechnology and medical science.[81] The patient's pre-social rights were nonetheless vindicated by finding that the patient was not afforded the opportunity to make informed choices as to his medical treatment. This judicial reasoning aligns with the contention that a person's *pre-social rights* in bodily material cannot include the right to profit from the transfer, relinquishment, or use of his or her bodily material.

If there is a right to profit from the provision of bodily material, the right must be premised upon a social basis. The second main argument has identified that such a social justification may be applicable to transfer or relinquishment of bodily material; allowing a progenitor (or executor) to profit from the provision of bodily material may enable the greater availability of bodily material for socially valuable enterprises. However, I have also suggested that this positive diachronic connection (between the right to profit from the relinquishment or transfer of bodily material and the wider distributional pattern of bodily material) will apply where there is, for a sufficient number of people, an equivalence between the value of the bodily material and the value of the financial benefit. The availability of the financial reward must then act as an extrinsic reason for relinquishing or transferring bodily

[81] *Moore v Regents of the University of California* (n 1) 143–4.

material that does not crowd-out concurring intrinsic reasons, and is able to defeat any intrinsic reasons for retaining control or possession of the bodily material.

In the instances where the financial benefit provides an extrinsic reason for action that is able to defeat intrinsic reasons for inaction, a risk of value denigration arises. The positive diachronic connection relies upon financial benefits to motivate the transfer of bodily material. The option to obtain a financial benefit can be transformed, for those in financially constraining circumstances, into social and financial pressure. Such pressure may force an equivalence between two non-equivalent things. That is, an equivalence between the value of the bodily material (that has value in its own right) and the substitutional value of the financial benefit. In these instances, the (first-order) value of the bodily material is denigrated and the (second-order) value of being self-ascriptive as to what about ourselves has value in its own right is being tainted. I have suggested that there are some decisions, such as what is inherently valuable to us or what is constitutive of us, that ought to be kept insulated from financial and social pressure. In terms of the use and control of bodily material, decisions as to whether bodily material is *for-itself* or *in-itself*, and decisions as to whether bodily material that is *for-itself* (or *for-others*) is retained or relinquished, are decisions that ought to be insulted from financial pressure.

4

The Concept of Property Law

In late 2004 Diane Blood and her husband started to try to conceive. In February 2005, her husband contracted meningitis. Upon Mrs Blood's request, samples of semen were extracted from her comatose husband. When Mrs Blood sought to use the semen for assisted reproductive therapy she was confronted by the statutory scheme (Human Fertilisation and Embryology Act 1990) and a regulatory authority (the Human Fertilisation and Embryology Authority). The Authority took the view that Mrs Blood's treatment would be contrary to statutory provisions that required 'written and effective consent of a man, after he has had the opportunity to receive counselling and after he has had a proper opportunity to consider the implications of a posthumous birth'.[1] The Authority therefore refused to authorize Mrs Blood's treatment.

Mrs Blood sought judicial review of the Authority's decision. In *ex parte Blood*, the Court of Appeal of England and Wales agreed that written consent for treatment is required for the use of stored gametes (although the court eventually quashed the Authority's decision on the grounds that it was inconsistent with her right to receive medical treatment in another European Union state).[2] The possession and use of the semen, in this case, was governed by the statutory provisions identified by the Authority (albeit in conjunction with European Community law).

In 2010, Mr and Ms Edwards sought help to conceive a child. The Edwards made an appointment with a fertility clinic to discuss their treatment options for 6 August 2010. On 5 August 2010, Mr Edwards was fatally injured in a work place accident. Ms Edwards applied for, and obtained, a court order that permitted the extraction of semen from the body of the late Mr Edwards. When Ms Edwards sought to use the semen she was also confronted by a statutory scheme (Assisted Reproductive Technology Act 2007).

[1] *R v Human Fertilisation and Embryology Authority, ex parte Blood* [1997] 2 All ER 687, 691.
[2] *R v Human Fertilisation and Embryology Authority, ex parte Blood* [1999] Fam 151.

The legislation prohibited the 'supply' of a 'gamete or an embryo to another person... except with the consent of the gamete provider'.[3]

In the intervening years between *ex parte Blood* and *Re: Edwards*, the common law had recognized property rights in semen.[4] In *Re: Edwards*, the Supreme Court of New South Wales held that Ms Edwards was entitled to possess the gametes, as items of property, by virtue of her being the administrator of the estate of her late husband, coupled with the finding that the semen was extracted 'for her purposes'.[5] According to the court, it then followed that the 'handing over of the stored gametes to her would more appropriately be regarded as them being "released" to her, rather than them being "supplied" to her'.[6]

In *ex parte Blood*, statute governed the use and storage of the semen. As a matter of private law, the Court of Appeal of England and Wales affirmed the regulatory authority's application of the statute. In *Re: Edwards*, the semen was as an item of property that Ms Edwards had the right to possess. It was this right to possession which enabled the Supreme Court of New South Wales to narrow the statutory restriction of the supply of gametes without the consent of the progenitor. The introduction of property rights in *Re: Edwards* enabled a novel legal analysis that differs in a number of ways from the analysis in *ex parte Blood*.

In this chapter we will explore how property law represents a particular way of creating legal relationships around the possession and use of an object or resource that enables a novel legal analysis. I will argue that property law is not merely the branch of law that governs the use of things, it is a branch of law that governs the use of things *in a legally distinct way*. Equipped with a conceptual understanding of property law, and coupled with a structural understanding of property law, developed in the next chapter, we are then in a position to assess whether entitlements in bodily material ought to give rise to property rights in bodily material.

I will explain how, for property law, the focal point of the legal relationship between the rights-holder and the duty-bearer is the object or resource itself. As a result, the law governs the ownership of an object or resource through the exclusion of others from the object or resource. Our relationship

[3] Assisted Reproductive Technology Act 2007 (NSW), s 21: 'An ART provider must not supply a gamete or an embryo to another person (including another ART provider) except with the consent of the gamete provider and in a manner that is consistent with the gamete provider's consent.'

[4] *Yearworth v North Bristol NHS Trust* [2009] EWCA Civ 37; *Bazley v Wesley Monash IVF Pty Ltd* [2010] QSC 118.

[5] *Jocelyn Edwards; Re the Estate of the Late Mark Edwards* [2011] NSWSC 478 [91] [emphasis added].

[6] *Re: Edwards* (n 5) [139].

with our property, and our relationship with our bodies, are both protected by the law through an 'exclusion strategy'. This 'exclusion strategy' contrasts with a 'governance strategy'. Under a governance strategy, the focal point of the legal relationship between the rights-holder and the duty-bearer is an activity or interaction that nonetheless may involve an object or resource. The law then governs the activity or interaction by allocating entitlements and responsibilities to participants in the activity or interaction.

The second conceptual feature of property law concerns the preferences and choices that the use of an object or resource enables. Property rights protect a set of preferences and choices that can exist independently of the rights-holder. One way of isolating whether a right is contingent to the rights-holder is to assess whether a subsequent rights-holder is able to stand in the same position with regards to the object or resource as the initial rights-holder. In comparison, non-contingent rights protect a set of preferences and choices that can only be exercised by the particular rights-holder. The preferences and choices are therefore dependent on, rather than contingent to, the rights-holder. I will argue that our rights of control and use over our bodies are non-contingent rights. Such rights of bodily integrity therefore conceptually differ from property rights. To conflate this distinction, between contingent and non-contingent rights, is to reduce a set of interests that have value in their own right to a set of interests that have value equivalence with the ownership of other things.

I. Property and Exclusion

I will explain in this section that a defining conceptual feature of a property right is the way in which the right excludes others from an object or resource. This conceptual feature can be unpacked with reference to three premises. These premises expressed together suggest that property rights are exclusive rights on the basis that:

(i) property rights protect our interest in purposely dealing with objects and resources;

(ii) which is an interest best conceived as an interest in an open-ended set of activities;

(iii) that the law protects through the exclusion of others from the object or resource.

In the next section, I will suggest that the same reasoning is equally applicable to the right to bodily integrity.

a. Property beyond Thing-ness

According to Laura Underkuffler, 'Property' is a volatile concept that is constructed by a 'complex package of normative choices'.[7] When assessing whether an object or resource ought to have the status of 'property' we ought not to assume that we all share a common understanding of 'property'. Rather, it is necessary to first clarify the social and moral assumptions about the use of an object or resource that the concept of property law is premised upon. Given a stable concept or idea of property law, we are then able to appreciate the structural or doctrinal features of property law (in the following chapter).

It is tempting to conceptualize property rights in terms of 'things-ness'. Property law would then be understood as 'the law that applies to things, objects or resources'. Land, water, furniture, toothbrushes, chickens, trademarks, kidneys, embryos, and spleen cells would all then be items of property. This basic functional definition of property rights as rights or entitlements 'in things, objects and resources' works if we are trying to justify why someone ought to be able to use, possess, control, or sell a thing, object, or resource.[8] In this chapter, and subsequent chapters, we have a different aim in mind. We are trying to isolate what it means for the ability to use, possess, control, or transfer a thing *to be protected* by property law, as compared to other branches of law.

The problem with the basic definition of property rights as 'rights in things' is that definition must presuppose a category of 'things' that is neither circular nor contestable. In other words, if property rights are rights in 'things', it remains to be explained what counts as a 'thing', and why. When we debate what 'things' count as items of property, the debate itself shifts our focus away from the 'thing-ness' of water, chickens, trademarks, and embryos to consideration of how property law characterizes the relationship between the rights-holder, the thing, and the duty-bearer. The discussion shifts beyond the thingness of the item of property to the content of the legally enforceable right. After all, 'property is not things but rights, rights in or to things'.[9] An

[7] LS Underkuffler, *The Idea of Property: Its Meaning and Power* (Oxford University Press, Oxford 2003) 11; A George, 'The Difficulty in Defining Property' (2005) 25 *Oxford J Legal Stud* 793, 795–6: 'Underkuffler criticises other scholars for tending to ignore the socially constructed and volatile nature of property or for acknowledging and then ending their inquires, and for assuming that their readers share a workable understanding of property.'

[8] JE Penner, *The Idea of Property in Law* (Oxford University Press, Oxford 1997) 106, *cf* J Waldron, *The Right to Private Property* (Clarendon Press, Oxford 1988) 33.

[9] CB Macpherson, 'The Meaning of Property' in CB Macpherson (ed) *Property: Mainstream and Critical Positions* (Basil Blackwell, Oxford 1978) 2.

adequate theoretical explanation of property law, one that can explain how property law is distinct from other branches of law, must therefore focus on legal relationship between between the rights-holder, the thing, and the duty-bearer.

Hence, I take a broad approach here. I assume that all branches of private law concern a rights-holder, a duty-bearer, a thing, and an activity. Where branches of law differ is how these basic features of law are organized, orientated, or configured in relation to each other. To start, I will contend that property rights are legal rights that focus on the exclusion of all other persons (other than the rights-holder) from an object or resource.

b. The Legal Relationship between the Rights-Holder and the Duty-Bearer

The 'thing-ness' view of property law has an intuitive appeal. It is true that property law is the area of law that governs the *use of things*. We assert our legal rights over our land, water, furniture, and toothbrushes because we have an interest in using these things. More generally, property rights represent our interest in 'purposely dealing with things'.[10] When a rights-holder has possession of an item of property, the possession entitlement *enables a set of activities* that the entitlement-holder may purposively engage in. For example, if Lucy possesses land (thing), she is able to grow wheat on the land (activity); if Matthew possesses a toothbrush (thing), he may brush his teeth with his toothbrush (activity), if Mark can access a body of water (thing), he may draw fish from the water (activity). The starting premise that explains the exclusivity of property law is that property law focuses on the thing rather than the activity. More precisely, for property law the focal point of the legal relationship between the rights-holder and duty-bearer is the *thing* (the land, the toothbrush, the water) rather than the activities that the use of the thing enables (the growing of wheat, the brushing of teeth, the harvesting of fish).

The alternative orientation is that the rights-holder has a legally protected interest in an action or activity. Although the activity may involve a thing, the activity remains the focus for the law. For instance, Lucy may contract with Joanna in order to obtain the right to grow wheat on Joanna's land, the manufacturer of toothpaste may owe Matthew a duty with regards to his oral health, and Mark may obtain a statutory licence to fish in a river. In these instances, the focal point of the law is an activity, although the activity nonetheless involves the use of things. This initial conceptual divide can therefore

[10] JE Penner, *The Idea of Property in Law* (n 8) 71.

be understood in terms of whether the thing, or the activity, is the focal point of the legal relationship between the rights-holder and duty-bearer.

c. An Open-Ended Set of Activities

Whilst property law protects our interest in 'purposefully dealing with things', James Penner observes that 'it is difficult in the extreme to quantify the many uses one can make of one's property'.[11] Lucy may grow barley or may graze cattle on her land, Matthew may clean his teeth, bathroom tiles, or his silverware with his toothbrush, Mark may farm fish in the river or create a nature reserve. Rather than cataloguing the possible ways in which an entitlement in an object or resource can be exercised, or listing the ways in which an object can be purposively dealt with, property law treats our *interest in things* as an interest in an *open-ended set of activities.* As James W Harris has remarked with regards to the use of land:

> The possessory owner has a *prima facie* privilege to do anything in relation to his land which the dominant culture of his society accords to a landowner. The set of privileges entailed is not a total set, because everywhere some things are excluded. It is, however, an *open-ended set,* since its present content could never be exhaustively listed.[12]

When the exercise of an entitlement is protected by a property right, the exercise of an entitlement may be pursuant to any number of possible activities, and such activities are presumptively permissible and protected from interference. Property law protects the exercise of an entitlement without identifying the set of activities that the exercise of an entitlement enables. In other words, the relationship between the rights-holder and the thing is not reduced to a set of activities that the thing enables. Rather, a property interest is an interest in a 'sphere of indefinite and undefined activity'.[13]

According to Thomas Merrill and Henry Smith's analysis of private law, open-ended entitlements are 'typical of ownership'.[14] Property rules, they argue, are used to protect entitlements for reasons of economic efficiency. For Smith, 'the preference for property rules can be understood as a response to

[11] JE Penner, *The Idea of Property in Law* (n 8) 71.

[12] JW Harris, *Property and Justice* (Clarendon Press, Oxford 1996) 160.

[13] JE Penner, *The Idea of Property in Law* (n 8) 72: 'The exclusion thesis is a statement of the driving analysis of property in legal systems. It characterizes property primarily as a protected sphere of indefinite and undefined activity, in which an owner may do anything with the things he owns.'

[14] L Katz, 'Exclusion and Exclusivity in Property Law' (2008) 58 *UTLJ* 275, 280; T Merrill and HE Smith, 'The Morality of Property' (2007) 45 *Wm & Mary L Rev* 1849; T Merrill and HE Smith, 'What Happened to Property in Law and Economics' (2001) 111 *Yale LJ* 257.

the information costs that shape other aspects of entitlement delineation'.[15] The 'exclusion strategy' of property law, according to Smith, 'relies on rough and low-cost signals that are not tied to use in order to protect indirectly a large and unspecified set of uses'. Property rules reduce information costs through the exclusion strategy by communicating the simple message of 'keep out' so that an undefined set of activities is protected without 'officials needing to know what these activities may be'.[16] It follows that 'owners' of property 'have open-ended choices of how to invest in or consume the asset'.[17] Mark knows not to use Matthew's toothbrush without Mark having to inquire into whether Matthew uses his toothbrush to clean his teeth, bathroom tiles, or his silverware; Joanna knows not to trespass onto Lucy's land without Joanna having to inquire into whether wheat seeds or barley seeds are germinating, or whether her cattle are calving or only grazing on the land.

In contrast to an 'exclusion strategy', a 'governance strategy' specifies valid and legally protected activities. A governance strategy is able to provide 'greater precision' in terms of the permitting and prohibiting activities (although it also introduces 'greater information costs that typically outweigh the benefits the greater precision governance rules provide').[18] For instance, given the competing and conflicting interest in the use of a river, statutory provisions or a network of contracts may be used to govern access to the river, the ability to discharge waste into the river, the ability to fish in the river, and so on. The law, in these instances, governs an identifiable set of activities.

Given the 'complexity of interactions between actors' each strategy, according to Smith, reduces interactions into 'modules' (or 'manageable chunks').[19] Property law and an exclusion strategy 'initially focuses on things', whereas tort law and a governance strategy 'takes more direct aim at acts and activities'.[20] The law of negligence and nuisance may, for instance, protect the activity

[15] HE Smith, 'Property and Property Rules' (2004) 79 NYU L Rev 1719, 1753: '[F]or reasons of information cost it is often advantageous and almost inevitable that rights will be delineated by means of what I have called an "exclusion strategy." Such a strategy relies on rough and low-cost signals that are not tied to use in order to protect indirectly a large and unspecified set of uses.'

[16] HE Smith, 'Property Rules' (n 15) 1728.

[17] HE Smith, 'Property Rules' (n 15) 1719.

[18] L Katz, 'Exclusion and Exclusivity in Property Law' (n 14) 280; T Merrill and HE Smith, 'The Morality of Property' (n 14) 1891.

[19] HE Smith, 'Modularity and Morality in the Law of Torts' (2011) 4 *J Tort L* 1, 2: 'Both property and torts solve the information cost problem with in rem rights in similar ways, by chopping up the world of interactions between parties into manageable chunks—modules—that are semi-autonomous.'

[20] HE Smith, 'Modularity and Morality in the Law of Torts' (n 19) 1: 'The main difference between property and torts is in their basic unit of analysis: property initially focuses on things and works out from there, whereas tort law takes more direct aim at acts and activities (which I will

of sowing seeds or brushing teeth, so that Lucy's interest in growing wheat is protected by duties on Lucy's neighbour, Joanna, restricting the use of herbicides. Matthew's interest in the activity of brushing his teeth is similarly protected by duties imposed on toothpaste suppliers to provide a safe product. Hence, whether it be because of the feature of our relationship with things, or as an information cost saving device, property law protects our interest is 'purposely dealing with things' which gives rise to a sphere of 'indefinite and undefined activity' where each entitlement is an open-ended entitlement.[21] In comparison, other branches of law 'take more direct aim at acts and activities'.[22]

d. The Exclusion of Others

In order for property law to protect an open-ended set of activities, it must provide a structure of rights and duties that protect an open-ended set of activities. Instead of positively identifying particular and preordained uses of property that are protected under the law, the 'contours' of a property right are provided by the exclusion of others from the object or resource.[23] The effect of focusing on the exclusion of others rather than the exercise of entitlements is to carve out a boundary around the sphere of 'indefinite and undefined activity'.[24] According to Larissa Katz, 'we might better characterise a boundary approach as a theory of non-ownership' since the contours of the property right are focused on excluding others from using the object or resource so that 'an owner is the last person standing after the exclusion of everyone else'.[25] Hence, in order to protect our interest in open-ended entitlements, the law constructs an exclusionary boundary of 'negative liberty'.[26]

For instance, Lucy may graze cattle on her land. If her neighbour (Joanna) were to carelessly cause the outbreak of a bovine disease, and thereby trigger

lump together as "actions"), in the sense that it focuses on conduct that potentially causes injury to others.'

[21] JE Penner, *The Idea of Property in Law* (n 8) 72.

[22] HE Smith, 'Modularity and Morality in the Law of Torts' (n 19) 1.

[23] JE Penner, *The Idea of Property in Law* (n 8) 72: 'The right to property is that normatively protected part of our interest in using property, and that part, i.e. that fraction of our uses of property, is determined by the extent to which others must exclude themselves from our property. Thus we are provided with the specific contours of the property right over the many different things that can be objects of property.'

[24] JE Penner, *The Idea of Property in Law* (n 8) 72.

[25] L Katz, 'Exclusion and Exclusivity in Property Law' (n 14) 277.

[26] JE Penner, *The Idea of Property in Law* (n 8) 73. *Cf* 'negative control' identified in *Yearworth v North Bristol NHS* by S Green, 'The Subject Matter of Conversion' [2010] *JBL* 218, 239: '[T]he Court gave much weight to the *negative* control of the claimants; that is, the fact that their sperm could neither continue to be stored, nor used for any purpose, without their consent.'

environmental restrictions on the movement of livestock, Joanna would be impairing (in a functional sense) Lucy's ability to 'use' her land. However, since the disease did not reach Lucy's land, Joanna would not be interfering with Lucy's property rights in her land.[27] Property law excludes Joanna from the land, but does not secure Lucy's right to freely move cattle on or off her land. 'What this illustrates', Simon Douglas explains, 'is that "use" is not a "right" because it is not directly protected by the law'.[28] This is where using Honoré's taxonomy of ownership begins to unravel. The taxonomy does not attempt to explain *the way in which* incidents of ownership attract legal protection. Despite 'the right to use' being a 'cardinal feature' of Honore's taxonomy, property law does not protect a positive right to 'use' an object or resource.[29] As we know from chapter one, applying the taxonomy therefore does not assist us in identifying the most applicable set of legal rules to protect the use of an object or resource.

Hence, the positive right to use an object or resource is not legally protected by property law per se. That said, the right to exclude others from an object or resource preserves a type of 'negative liberty' that enables, or preserves, a set of uses and activities.[30] The first premise (that property law governs the use of things) and the second premise (that property law excludes all others from the thing) are therefore correlative. As Penner explains:

> [I]n the social setting in which we live, and the ways in which things of this world are typically used, we see that any meaningful right to use is the opposite side of the coin to the right to exclude.[31]

Property rights construct a boundary to protect the sphere of an undefined, or open-ended set of, activities that gives ascendancy to the right-holder's interests, preferences, and choices to the exclusion of others. The legal boundary is constructed around the item or property itself, and this allows 'strangers to interact with each other in a rule governed way, though their dealings are not personal in any respect'.[32] Property, as Charles Reich describes:

[27] *D Pride & Partners v Institute for Animal Health* [2009] EWHC 685 (QB).

[28] S Douglas, 'Property Rights in Human Biological Material' in I Goold and others (eds), *Persons, Parts and Property: How Should We Regulate Human Tissue in the 21st Century?* (Hart Publishing, Oxford 2014) 89, 105–6.

[29] S Douglas, Property Rights in Human Biological Material', (n 28) 105; AM Honoré, 'Ownership' in *Making Law Bind: Essays Legal and Philosophical* (Clarendon Press, Oxford 1961) 116.

[30] JE Penner, *The Idea of Property in Law* (n 8) 73.

[31] JE Penner, *The Idea of Property in Law* (n 8) 71.

[32] JE Penner, *The Idea of Property In Law* (n 8) 30.

draws a circle around the activities of each private individual or organization. Within that circle, the owner has a greater degree of freedom than without. Outside he must justify or explain his actions, and show his authority. Within he is master and the state must justify any interference.[33]

A consequence of this 'exclusionary boundary' is a basic *inequality* between the rights-holder and the duty-bearer. The rights-holder, within the boundary, may engage in an undefined set of activities, whereas, the duty-holder, outside the boundary is excluded from any activity with regards to the thing. In contrast, where the law employs the governance strategy and focuses on actions and activity, a 'bilateral structure'[34] apportions rights and responsibilities between participants in an action or activity from a starting point of 'basic equality'.

If we are to characterize property rights as *exclusive* rights, let us be careful as to what the right to exclude entails in this context. It is not the preclusion of anyone else from having entitlements in the property. Property law, for instance, recognizes subsidiary and reversionary interest holders.[35] Rather, the right to exclude is the preservation of the owner's position of the exclusive 'agenda-setting authority'[36] which is a 'position that is neither derived from nor subordinate to the position of others with respect to that resource'.[37] To that extent, 'the right to property is like a gate, not a wall'; it permits the rights-holder 'to make a social use of his property, by selectively excluding others, which is to say by selectively allowing some to enter'.[38]

From the starting point that all branches of private law concern the use of an object or resource in performance of an action or activity by the rights-holder, I have suggested that the focus of property law is on the object or resource itself. This focus permits the rights-holder to engage in an open set of activities. Property law then functions to protect this sphere of undefined activity through the exclusion of all others from the object or resource. The

[33] C Reich, 'The New Property' (1964) 73 *Yale LJ* 733, 739.

[34] HE Smith, 'Modularity and Morality in the Law of Torts' (n 19) 17: 'One of the most striking features of tort law . . . is its bilateral structure. Tort claims run between a right holder and a duty bearer.'

[35] L Katz, 'Exclusion and Exclusivity in Property Law' (n 14) 275: 'So long as others—whether they be holders of subsidiary property rights or strangers to the property—act in a way that is consistent with the owner's agenda, they pose no threat to the owner's exclusive position as agenda setter.'

[36] L Katz, 'Exclusion and Exclusivity in Property Law' (n 14) 289.

[37] L Katz, 'Exclusion and Exclusivity in Property Law' (n 14) 297, 315.

[38] JE Penner, *The Idea of Property in Law* (n 8) 74: 'The right to property is like a gate, not a wall. The right to property permits the owner not only to make solitary use of his property, by excluding all others, but also permits him to make a social use of his property, by selectively excluding others, which is to say by selectively allowing some to enter.'

alternative orientation places the activity undertaken by the rights-holder, or interaction between the rights-holder and duty-bearer, as the focal point. The law then governs this interaction by allocating entitlements and responsibilities between the parties to the interaction. The *exclusion of others* is a conceptual feature of property law, and it can be used to explain (in the next chapter) a number of important structural features of property law, such as: why property rights are exercisable against an open set of persons, why interference with property rights is actionable per se and why property rights impose duties of non-interference. Yet, as we shall now turn our attention to, the *right to exclude* is not the sole preserve of property law.

II. Bodies and Exclusion

Property rights are not the only exclusive rights in the legal system. Our right to bodily integrity, which gives legal force to the principle of self-ownership, is also an exclusive right. As Smith notes,[39] 'tort law treats the acts and activities surrounding bodily integrity and reputation in [a way] which bears some resemblance to the ... exclusion strategy in property'. I will suggest here that our rights of use and control over our bodies also represent a sphere of undefined activity that is protected by an exclusionary boundary. I will demonstrate this by retracing the premises that explain property law's exclusionary boundary.

Recall that the principle of self-ownership confers, according to Gerald Cohen, 'the fullest right a person (logically) can have over herself provided that each other person also has just such a right'.[40] As we have seen in chapter two, there are various different ways of justifying the self-ownership of the body. The rights of control and use may be necessary preconditions for autonomous action, and may provide the most efficient means of maximizing welfare, utility, or preference satisfaction. These bases of self-ownership explain why we have an interest in 'purposely dealing with'[41] our bodies. We purposely use and control our bodies as the primary and most immediate way in which we exercise our autonomy, liberty, or preference satisfaction (or however we understand the subjectivity of the person). It is because we have such a direct and significant interest in the control and use of our bodies that

[39] HE Smith, 'Modularity and Morality in the Law of Torts' (n 19) 16.

[40] GA Cohen, *Self-Ownership, Freedom and Equality* (Cambridge University Press, Cambridge 1995) 213.

[41] JE Penner, *The Idea of Property in Law* (n 8) 71.

our bodies become a focal point of the legal relationship between ourselves and others.

The set of activities that the self-owner may engage in is therefore an open-ended set of activities. This is because the body, in pursuit of various different personal aims, objectives, standards, and values, can be used in innumerable ways. The relationship between a person and his or her body is thus not reducible to a set of activities that the body enables. For example, Matthew may consent to physical contact by playing rugby, to the physical intrusion of a medical procedure, or to physical contact in sexual intercourse, or he may not consent to any of these activities. All of these activities, and many others, are shaped around Matthew's rights of use and control of his body. How these rights of use and control are exercised cannot be 'exhaustively listed'.[42] Instead, the self-owner 'has a *prima facie* privilege to do anything in relation' to his or her body 'which the dominant culture of his society accords' to the self-owner.[43]

Akin to the use and control of property, since it is difficult to completely audit the many uses one can make of one's body, the best way to conceive of the principle of self-ownership is as a sphere of presumptively permitted activity. Although limits may be prescribed as to what activities are impermissible, the presumption remains that as between a person and their own body, a person's self-ownership, and their legal right to bodily integrity, represents a sphere of undefined activity.

The way in which the law then protects these open-ended entitlements in the body is by focusing the legal attention on sources of interference rather than the exercise of entitlements of use and control. The contours of the right to bodily integrity are provided by the conduct of others that is prohibited. Since the body exists in a social context that involves interactions with a range of other people, in order to protect the open-ended entitlements in the body, the law excludes others from the body. To borrow Katz's characterisation, it is a theory of non-ownership—or negative ownership—insofar as the self-owner is the last person standing following the exclusion of everyone else.[44] The right to exclude others does not forgo the ability to include others. Consent provides the 'gate' in the exclusionary boundary that allows others to tackle,[45] cut into, or touch, a self-owner's body, and it is the requirement of consent that preserves the self-owner's 'agenda-setting authority'[46] over the body.

[42] JW Harris, *Property and Justice* (n 12) 160.
[43] JW Harris, *Property and Justice* (n 12) 160.
[44] L Katz, 'Exclusion and Exclusivity in Property Law' (n 14) 277.
[45] JE Penner, *The Idea of Property in Law* (n 8) 74.
[46] L Katz, 'Exclusion and Exclusivity in Property Law' (n 14) 289.

The rights of bodily integrity that vindicate the principle of self-ownership are, therefore, exclusionary rights. Similar to Reich's conception of property,[47] outside the exclusionary boundary, 'one must justify or explain his actions' to justify the interference with the bodily sphere that is presumptively impermissible. Inside the boundary, the exercise of entitlements is presumptively valid and protected since the self-owner exercises his or her autonomy and preferences through the exercise of entitlements in his or her body. The body is a social object that interacts with others on a daily basis. The requirement of consent for any interaction with, or intrusion into, the bodily sphere represents a 'gate' through which we can selectively allow some to interact with the body whilst 'selectively exclud[ing] others'.[48]

This picture, of exercising an authority over the body that is an exclusive authority insofar as it 'is neither derived from nor subordinate to the position of others',[49] portrays the interests we have, or ought to have, in our own body. Property is a more familiar instance of exclusive rights. This explains why describing the *body-as-property* has a persuasive appeal. By pulling rights of the body into the established exclusionary boundaries of property law, 'property' becomes 'the ideal symbol' for personal autonomy 'for it [can] both literally and figuratively provide the necessary walls to separate oneself from others'.[50] It also explains the intuitive appeal behind applying property law to the entitlements that we may have in bodily material. As subsequent chapters will explain, there are many instances where our entitlements in bodily material ought to be protected by an exclusionary boundary.

Property rights and rights of bodily integrity therefore share a conceptual feature. They are both principally concerned with the purposeful use of things, to the extent that entitlements in the body and property enable an open-ended set of activities. As a result, the law protects these open-ended entitlements by focusing on the potential sources of interference with the entitlement. Property rights and rights of bodily integrity are both therefore exclusive rights.

As we will see in the next chapter, it is this conceptual feature of property law that explains the motivation to treat items of bodily material as items of property: the legal structure of the exclusionary boundary provides robust protection of entitlements in things. Yet, as I will also explain, since property rights are not the only form of exclusive rights, it is possible to provide exclusionary protection without the application of property law. The reason

[47] C Reich, 'The New Property' (n 33) 739.

[48] JE Penner, *The Idea of Property in Law* (n 8) 74.

[49] L Katz, 'Exclusion and Exclusivity in Property Law' (n 14) 297.

[50] J Nedelsky, 'Law, Boundaries and the Bounded Self' (1990) 30 *Representations* 162, 168.

why we ought to consider alternatives to property law stems from a further conceptual feature of property law. We shall now turn to consider this further feature.

III. Property and Contingency

Although there is a conceptual similarity between property rights and rights of bodily integrity, there is also an important conceptual distinction between these branches of law. This section aims to identify a further conceptual feature of property law that differentiates property rights from other exclusive rights. I will suggest here that property rights enable preferences and choices that can exist independently of the rights-holder.

a. Munzer's Criterion of Transferability

For Stephen Munzer, the 'most useful criterion' for distinguishing 'property rights' from 'personal rights' (such as free speech, privacy, reputation, and bodily integrity) is 'transferability'.[51] If rights are conferred on an entity to protect their interests and choices, then according to Munzer:

> Personal rights are … rights that protect interests or choices other than the choice to transfer. Property rights are rights that protect the choice to transfer.[52]

Within the concept of property rights there is, for Munzer, a distinction between weak property rights, which protect gratuitous transfers, and strong property rights, which protect transfer for value.[53] Hence, Munzer's distinction between degrees of property maps onto JW Harris' ownership spectrum (discussed at pages 30–1).[54] Once a collection of entitlements moves along the spectrum, past control entitlements to include transfer entitlements, 'mere property' arises. Moving further along the ownership spectrum to include income entitlements, the owned-thing becomes (strong) property.

Munzer is correct that property rights are transferable and personal rights (such as bodily integrity, privacy, and reputation) are non-transferable. However, this distinction as to what can be done with an entitlement is a structural or doctrinal distinction (which we will consider in the next chapter). We need to first isolate the deeper, conceptual, distinction that is able

[51] SR Munzer, *A Theory of Property* (Cambridge University Press, Cambridge 1990) 47.
[52] SR Munzer, *A Theory of Property* (n 51) 48.
[53] SR Munzer, *A Theory of Property* (n 51) 49.
[54] JW Harris, *Property and Justice* (n 12) 27.

to explain *why* property rights are transferable and *why* personal rights are non-transferable.

b. Two Criteria for Contingency

Property rights and the right to bodily integrity protect the exercise of entitlements in items of property and the body respectively. The exercise of entitlements in things, such as items of property, personal information, or the body, enables a set of preferences and choices. For instance, Joanna may have exclusive use and control over her wine collection, a rare impressionist painting, the results of her blood test and, of course, exclusive use and control over her body. The wine enables her to enhance a culinary experience, the painting provides a stunning visual centre piece to her living room, the blood results enable her to make informed and private decisions about her medical treatment, and her body—first and foremost—enables the experience of these things (and the experience of all other things).

The way in which our rights over our property and over our body differ is in terms of the connection between the rights-holder and the preferences and choices that the exercise of entitlements enables. The use and control of an item of property enables preferences and choices that can exist independently of the rights-holder. Joanna's wine and her painting enable preferences and choices that are only contingently associated; such preferences and choices are not associated with her in any particular or necessary way. In comparison, personal rights enable preferences and choices that are necessarily associated with the rights-holder. The preferences and choices enabled by the exclusive control over blood test results and her exclusive control over her body are preferences and choices that can only be exercised by Joanna.

This characterisation of property rights as contingent rights follows from Penner's 'Separability Thesis', which suggests that:[55]

> Only those 'things' in the world which are contingently associated with a particular owner may be objects of property.

Another way of formulating the distinction is to suggest that:

> The essential feature of property is that it has an existence independent of a particular person.[56]

[55] JE Penner, *The Idea of Property in Law* (n 8) 111.

[56] *OBG v Allan* [2007] UKHL 21 (Baroness Hale) [309]: 'The essential feature of property is that it has an existence independent of a particular person: it can be bought and sold, given and received, bequeathed and inherited, pledged or seized to secure debts, acquired (in the olden days) by a husband on marrying its owner.'

In order to isolate this distinction between contingent and non-contingent rights, we will consider a set of criteria. The first criterion concerns the normative consequences that follow from the change in rights-holder. According to Penner, if a right is contingently associated with a rights-holder, then:

> in theory nothing of normative consequence beyond the fact that the ownership has changed occurs when an object of property is alienated to another.[57]

When the control of an object or resource is transferred to another person there will be, in all cases, normative consequences insofar as entitlements have changed hands. These are 'mere allocative consequences'. According to the Separability Thesis, where an item of property is transferred, the only normative consequence is the shift in entitlements from the transferor to the transferee. There are no normative consequences (beyond mere allocative consequences) to the transfer of property.

One way to assess whether there are normative consequences (beyond mere allocative consequences) as a result of the change of ownership is to ask whether 'a different person who takes on the relationship to the thing stands in essentially the same position as the first person'.[58] Hence, according to the second criterion, if a subsequent rights-holder is able to stand in 'the same position' with regards to the object or resource as the original rights-holder, then the relationship between the original rights-holder and the object or resource is a contingent relationship.

What these two criteria are able to isolate is whether the preferences and choices that the exercise of an entitlement in a thing enables are preferences and choices that can exist independently of the entitlement-holder. If such preferences and choices can exist independently of the entitlement-holder, then another person can stand in the same position as the original entitlement-holder and there will be no normative consequences (beyond mere allocative consequences) from the change in entitlement-holder. However, if the preferences and choices cannot exist independently of the entitlement-holder, then another person cannot stand in the same position as the original entitlement-holder and there will be normative consequences (beyond mere allocative consequences) were someone to acquire or interfere with the entitlements of the entitlement-holder.

Let us return to Joanna's collection of exclusive rights. If Lucy were to obtain Joanna's blood test results or infringe Joanna's bodily integrity, there would be (in terms of the first criterion) normative consequences *beyond* Lucy

[57] JE Penner, *The Idea of Property in Law* (n 8) 111.
[58] JE Penner, *The Idea of Property in Law* (n 8) 113.

obtaining possession of the test results and Lucy exerting physical control over Joanna's body. Through the invasion of her privacy and infringement of her bodily integrity Joanna suffers a loss—a normative consequence—that does not equate with something that Lucy gains. The reason why Joanna's loss cannot be equated with Lucy's gain is because (in terms of the second criterion) Lucy cannot stand in the same position as Joanna with regards to her privacy or bodily integrity. This is because the relationship between Joanna and her interests in her bodily integrity and privacy is a relationship that is particular to her; it cannot be substituted nor can another person adopt the relationship. Hence, the preferences and choices that Joanna's relationship with her body and her personal information enable are preferences and choices that cannot exist independently of her.

In contrast, if Lucy were to obtain Joanna's wine collection or rare impressionist painting, there would be no normative consequences beyond the (lawful or unlawful) change in entitlement-holder. Apart from shifting the entitlements in the wine and painting from Joanna to Lucy (mere allocative consequences), there are no normative consequences of depriving Joanna of her preferences in, and choices regarding, her wine and painting. Lucy would gain what Joanna would lose. This normative symmetry is because Lucy is able to acquire and exercise the same preferences and choices regarding the wine and painting that Joanna had prior to her. In other words, the *content* of the ownership relationship between the rights-holder and the thing is the same because the content of the ownership relationship is contingent to the rights-holder. The relationship between Joanna and her wine and painting is a contingent relationship; it can be substituted for other goods and another person can adopt the same relationship with the goods.

Note that the 'separability' ('contingency' or 'independence') of property is the separability of the *right* (rather than the separability of thing) from the rights-holder. Again, 'property is not things but rights'.[59] Some have misunderstood the separability or contingency of property rights and, when applying the conceptual distinction to bodily material, have prescribed unnecessary threshold requirements of 'substantial changes to constitutive parts of our personality' or complete deprivation 'of our status as persons'.[60] The direction of the contingency or independence has been misinterpreted. It is not that our personalities depend on non-contingent rights (so that the deprivation of a non-contingent right is a drastic event for our 'personalities' or 'status as

[59] CB Macpherson, 'The Meaning of Property' (n 9) 2.

[60] M Quigley, 'Property in Human Biomaterials—Separating Persons and Things?' (2012) 32 *Oxford J Legal Stud* 659, 672.

persons'). Rather, the existence of non-contingent rights is dependent on the rights-holder (so that the change in rights-holder would be fatal to the existence and content of the right). In short, Lucy's right to privacy is dependent upon Lucy's existence (as the rights-holder), her existence is not dependent upon her right to privacy!

The distinction between contingent rights and non-contingent rights ultimately depends on the types of preferences and choices that the possession, use, or control of a thing enables. We can isolate this conceptual distinction by isolating two related criteria. We can first ask whether there are further or additional 'normative consequences from [the] change in rights-holder' (beyond mere allocative consequences). We can then assess whether there are normative consequences from the change in rights-holder by asking whether the subsequent possessor, user, or controller of the thing being is able to 'stand in the same position' with regards to the thing as the original entitlement-holder. To be unable to 'stand in the same position as' another person with regards to an object or resource is shorthand for being unable to acquire and exercise the same preferences and choices that the object or resource enables for the initial rights-holder.

Hence, where the exercise of entitlements in an object or resource gives rise to preferences and choices that are 'contingently associated with a particular owner'[61] or exist 'independent of a particular person',[62] such entitlements in an object or resource represent contingent rights. All property rights fall into this conceptual category. In comparison, if the exercise of entitlements in an object or resource gives rise to preferences and choices that are necessarily associated with a particular person or cannot exist independently of the particular person, then such entitlements represent non-contingent rights. Our 'personal rights', such as our right to bodily integrity, privacy, and reputation, are non-contingent rights.

c. Returning to the Hegelian Blur

This conceptual division between contingent (property) rights and non-contingent (personal) rights may not appear to be a clean division. It is arguable that there are some ownership relationships that do not easily fit into either category. If there are 'contestable cases of property', then perhaps some 'personal rights' can also fall into this intermediate category. In particular, perhaps the rights that arise in bodily material may fall into the intermediate category and represent some kind of weak form of a property

[61] JE Penner, *The Idea of Property in Law* (n 8) 111. [62] *OBG v Allan* (n 56) [309].

right. Against this suggestion, I will argue here that the criteria of contingency does not allow for an intermediate category, even where the distinction between object and subject is blurred, such as when the distinction is blurred by Hegel's 'nexus of dual relations' (see pages 50–2).

Consider, for example, Margaret Radin's 'property for personhood'. These are items of property that are characterized by 'coexistent commodified and non-commodified understandings' which give rise to 'incomplete commodities'.[63] Radin uses housing and employment as examples of incomplete commodification:[64] Our home is 'a sanctuary needed for personhood'[65] and our employment is, according to Radin:

> not only the way we make our living, but it is also part of ourselves. What we hope to derive from our work, and the personal importance we attach to it, are not understandable entirely in monetary terms, even though we demand and accept money. The ideals about work seem to be part of our conception of human flourishing.[66]

Radin is correct that there are incomplete commodities insofar as our housing or employment engage a set of values that other commodities may not. Yet, these 'contestable' cases that Radin presents nonetheless represent cases of contingent rights. Although Lucy may see her personhood or subjectivity expressed in an external object (a house, a wedding ring, a position of employment), it is equally possible for Joanna to have entered into, or to subsequently enter into, the same relationship with the house, wedding ring, or position of employment. Although the transition may be more complex than with other objects or commodities, Joanna can leave her job, sell her ring and home, and Lucy can obtain her job, buy her ring and her home. In such cases, although the ring, job, or home are 'vehicles' or 'sanctuaries' for personhood, they are nonetheless *contingent* vehicles or sanctuaries for personhood; another person can stand in the same position with regards to the commodity than the prior entitlement-holder.

Radin is able to complicate the distinction between property and personhood because her explanation for why we ought to be entitled to these 'contested commodities' is a Hegelian explanation. As we know from chapter two, the Hegelian explanation blurs or complicates the distinction between object and subject, and it is this ambiguity that the Radin account

[63] MJ Radin, *Contested Commodities* (Harvard University Press, Cambridge MA 1996) xiii; MJ Radin, 'Property and Personhood' (1982) 34 *Stan L Rev* 957; CH Harrison, 'Neither Moore nor the Market: Alternative Models for Compensating Contributors of Human Tissue' (2002) 28 *AJLM* 77, 90.

[64] MJ Radin, 'Market Inalienabilty' (1987) 100 Harv L Rev 1849, 1918.

[65] MJ Radin, 'Property and Personhood' (n 63) 995.

[66] MJ Radin, 'Market Inalienability' (n 64) 191–9.

of contested commodities is able to trade on. Recall that the Hegelian basis for the exercise of entitlements in objects or resources follows from the extension of the 'inherently abstract will' into the external sphere. According to Radin, the extension of the subject is able to 'bridge the gap' between object and subject but it does so in a way that blurs 'the boundary, between the self and the world, between what is inside and outside, between what is subject and object'.[67] The implication is the Hegelian explanation blurs the distinction between the subject and object. Contested commodities are therefore items of property that fall somewhere along this blurred distinction.

On one hand, it is possible to interpret the Hegelian relationship between the subject and the external object as a relationship that is 'constitutive' of the subject. As Ferguson explains, the Hegelian subject 'always has to go outside itself to know what is inside; by seeing itself reflected in the world it discovers relations *constitutive* of itself'.[68] If the relationship between a subject and an object is constitutive of the subject then the relationship represents preferences and choices that are particular or necessary to the subject themselves. Hence, there may be 'noncommodified understandings' of items of property, and such items of property may fall within the sphere of a person's subjectivity.

On the other hand, Jeremy Waldron interprets the Hegelian explanation as creating a relationship between a subject and object that ought to attract our duties of respect whilst also being a relationship that is both alienable and commercial.[69] Under this interpretation of the Hegelian explanation, the 'dual nexus of relations' between the subject and object justifies the right to private property even where the rights of the property rights-holder include the right to alienate or sell the object.

Unless we interpret the Hegelian account as forming relationships with objects that are constitutive of the subject, the Hegelian explanation risks a paradox. According to Margaret Davies and Ngaire Naffine:

> The paradox within Radin's work is that it sets property against property. The self is understood as a function of property, and this propertised self is in turn expected to protect against the commodity form of the person.[70]

[67] MJ Radin, 'The Colin Raugh Thomas O'Fallan Memorial Lecture on Reconsidering Personhood' (1995) *Or L Rev* 423, 426; MJ Davies and N Naffine, *Are Persons Property? Legal Debates about Property and Personality* (Ashgate, Aldershot 2001) 7.

[68] KE Ferguson, *The Man Question: Visions of Subjectivity in Feminist Theory* (University of California Press, Berkeley 1993) 41 [emphasis added].

[69] J Waldron, *The Right to Private Property* (n 8) 343–423; JW Harris, *Property and Justice* (n 12) 245.

[70] MJ Davies and N Naffine, *Are Persons Property* (n 67) 9.

If an item of property embodies the expression of the subject into the 'external sphere' to the extent that it represents preferences and choices that can only be exercised by the subject, then it is a curious feature of Radin's work that she uses property law to fight against the further alienation or commodification of a person's subjectivity or personhood.

More broadly, the central difficulty with applying the Hegelian account of *property rights* is being able to describe a relationship that is simultaneously sufficiently close whilst also being sufficiently separable. The Hegelian account justifies entitlements in things by showing a sufficiently close link between the subjectivity or personhood of person and the object so as to justify why the subject ought to have entitlements in the object. At the same time, given the conceptual contingency of property rights, the Hegelian account of property must also explain how the relationship between subject and object is sufficiently contingent to the subject so that the object can be treated as an item of property. We can avoid this paradox if we, following the analysis in chapter one, accept the validity of the Hegelian account of *ownership* whilst resisting the presumption the Hegelian basis ought to give rise to *property* rights.

That is not to deny that the division between subject and object may be a blurred division. A range of things may be both vehicles and sanctuaries for personhood, whilst nonetheless being ultimately contingent to the person. To the extent that the structure of the law operates on the basis of ultimately blunt divisions between categories of rights and branches of law, our focus remains on the conceptual divide between contingent and non-contingent rights. We shall now turn to consider whether our rights of exclusive use and control over our body are contingent or non-contingent rights by applying the criteria developed in this section.

IV. Bodies and Non-Contingency

When we consider the criteria for isolating contingent rights it would appear that our right to bodily integrity is a right that is necessarily associated with us. This is because our use and control of our body appears to enable preferences and choices that can only be exercised by us. For instance, my choice to consent to, or refrain from, physical contact and the application of force through a contact sport, an intimate relationship, or a medical procedure represents a choice that only I can exercise. Another person cannot stand in the same position as me with regards to the choices and preferences that I exercise and satisfy through the use of my body. Here we will consider why, and to what extent, the preferences and choices that underlie our exclusive use and control of our bodies are preferences and choices that cannot exist independently of us. It is crucial that we assess the extent to which our rights of use and control over our bodies are necessarily associated with us as the self-owner. This is because how we view the legal relationship between a person and his or her body will inform our view of the legal relationship between a person and his or her bodily material, and we are ultimately interested in whether the latter relationship is a contingent or non-contingent relationship.

However, whether our use and control of our body are rights that are separable or inseparable from the person will depend on how we view the relationship between a person's subjectivity (the features or characteristics of a person that ground moral duties towards them) and a person's material or physical body. In chapter two I compared two ways to view the relationship between a person's subjectivity and their body. It is now necessary to consider whether, according to these views, our exclusive rights of use and control over our bodies are rights that are contingently associated with us, or necessarily associated with us, as the rights-holder.

a. Contingency and the Third-Person Point of View

According to the third-person point of view of the body, the body is a set of complex physiological systems that are causally connected with the mind or subjectivity of the person. Simply put, the body is the 'chunk of the material world that sits in closest contact with [the] mind'.[71] The close causal connection between the subjectivity of the person and his or her body is the basis for the principle of self-ownership (and the legal right to bodily integrity). Since the body is the most immediate way in which a person executes rationally motivated action, or satisfies that person's preferences, wants or desires, the body performs a key role in enabling a person's autonomy, welfare, or preference satisfaction. A person's rights of exclusive use and control over their body are justified as a means of promoting the autonomy, welfare, or preference satisfaction of the person. According to this point of view, the body has value only as a means of achieving these further ends of autonomous action, welfare, or preference satisfaction (or however a person's subjectivity is constituted).

Although 'the structures and capacities of the body are contingent and arbitrary facts about the kinds of creatures we happen to be',[72] the body is

[71] T Carman, *Merleau-Ponty* (The Routledge Philosophers, Routledge, 2008) 82.
[72] T Carman, *Merleau-Ponty* (n 71) 81.

nonetheless necessarily associated with the person *to an extent*. It is a necessary association only insofar as, at any fixed point in time, another person cannot stand in the same position with regards to the body as the self-owner because another person cannot obtain or acquire the close or immediate causal connection between a person's subjectivity and the physical body. The necessary connection between a self-owner and their preferences and choices, that is facilitated through the right to bodily integrity, is sustained (only) by the close or immediate connection between his or her subjectivity and his or her body.

When we consider whether there are normative consequences (beyond mere allocative consequences) from the change in self-owner, we start to appreciate how the necessary association between a person and his or her body may only *temporarily* be a necessary association. If the body has value insofar as it enables a person's autonomy, welfare, or preference satisfaction, and if the body can perform the same task for another person, then the only normative consequences would be allocative consequences in terms of the substitution of a person's body for another. Organ transplants provide a straightforward example. The recipient of a kidney, for instance, is able to stand in the same position with regards to the kidney as the provider of the kidney. The kidney is able to, for the recipient, differentiate blood cells, minerals, and waste materials to produce urine in a functionally equivalent way as it did for the provider of the kidney. The kidney only has value in terms of this function, and this function is substitutional between persons. The implication of the third-person point of view is that the body, as a set of 'contingent and arbitrary facts' about the person, may also be entirely substitutional in this way.

As a result, if we adhere to the third-person point of view, the necessary relationship between a person and their body is an error that is carried forward from our first-person experience of the body. Our body and its parts may, one day, be entirely substitutional and contingent facts about ourselves. As we can anticipate, if our bodies are only contingently associated with us, then our relationship with an item of separated bodily material will also be a contingent relationship. Moreover, once separated, it is no longer immediately connected with our autonomy, welfare, or preference satisfaction. It would follow that any interest we may have in an item of bodily material is conceptually consistent with a property interest, and therefore property law is likely to provide the most appropriate structure of legal protection for entitlements in bodily material.

b. Contingency and the First-Person Point of View

In chapter two I outlined the view that our subjectivity is a physical subjectivity.[73] One of the implications of this view is that our subjectivity is interwoven with the physical environment. Our physical environment, prior to anything else, is our material and physical bodies. If we understand our subjectivity in terms of 'tasks' (the integration of components of our subjectivity with the physical environment) then our subjectivity is located in the body insofar as the body is 'polarised by its tasks, it exists towards them, it gathers itself up to reach its goal'.[74] Rather than there being a close connection between the body and subject, the body and subject are integrated or even 'identical' insofar as we cannot separate our tasks (or our subjectivity) from the body.[75]

In terms of the first criterion of contingency, no other person can stand in the same position as ourselves with regards to our body because our body is always 'directed towards a certain existing or possible task' or project.[76] Each existing and possible task represents the unity or integration of components of our subjectivity (our standards, goals, preferences, pain, or pleasurable experiences) with our physical and material environment. The body, first and foremost, is the location or site of this integration. It is not possible, therefore, for another person to take on the same position as ourselves with regards to our body. In terms of the second criterion of contingency, there would be normative consequences (beyond mere allocative consequences) if someone was able to be substituted into the body of another since the standards, goals, preferences, pain, and pleasure of the initial self-owner are not substitutable.

The body also has a further value beyond being directed towards tasks (being *for-itself*). The body is also *for-others*. Although we may admire the intentions, decisions, and perceptions of others (their 'inner subjectivity'), the way in which these characteristics and attributes are exhibited is necessarily through their bodies. The subjectivity of others, as perceived by us, is a physical subjectivity. We interact or engage with another through the physical-material environment. The relationship between two people is therefore a relationship between two embodied subjects. It follows that our relationship with the body of another person is a relationship that is particular

[73] S Priest, *Merleau-Ponty* (Arguments of the Philosophers Series, Routledge, London 1998) 57.
[74] M Merleau-Ponty, *Phenomenology of Perception*, C Smith (tr), (Routledge, London 2003) 115.
[75] T Carman, *Merleau-Ponty* (n 71) 81.
[76] M Merleau-Ponty, *Phenomenology of Perception* (n 74) 100; SK Toombs, 'What Does it Mean to Be Somebody?: Phenomenological Reflections and Ethical Quandaries' in MJ Cherry (ed) *Persons and Their Bodies: Rights, Responsibilities, Relationships* (Kluwer Academic Publishers, Dordrecht 1999) 81.

to us. We engage or interact with another person in our capacity as a parent, as a child, as a spouse, as a sibling, or as a friend. Another person, therefore, cannot stand in the same position as us with regards to the body of another since they cannot acquire our role or capacity as a parent, as a child, as a spouse, as a sibling, or as a friend.

Take, for example, Matthew, who has undertaken a project of learning to play Rachmaninoff's third piano concerto. He also enjoys drinking wine regularly, has the goal of playing first-class cricket, and has an interest or desire in sexual intimacy with his partner. These projects, preferences, goals, and interests form part of Matthew's subjectivity. Consider a scenario where Matthew undertakes a full body transplant (or even a significant interchange of organs and body parts) with Mark. Matthew would not be able to pursue his projects, preferences, goals, or interests. In terms of Rachmaninoff's third piano concerto, his new body would not have the 'muscular understanding' to manoeuvre his fingers across the keys. For Merleau-Ponty, 'it is the body that understands in the acquisition of habit'.[77] The idea is that, through habitual training, the body becomes increasingly responsive and expert at certain tasks. To that extent, Matthew's goal of playing first-class cricket has also vanished following his body transplant. With new taste buds, even Matthew's enthusiasm for wine may change or at least need to be more restrained if his new kidneys struggle to cope with the toxins from the alcohol. Since Matthew's body is also *for-others*, Matthew's partner is much less likely to engage in sexual activity with Matthew (now in Mark's body).

Perhaps, over time, Matthew could acquire the dexterity required for cricket and concerti. Perhaps Matthew's partner could learn adjust to Matthew's new appearance, and perhaps Matthew could develop an appreciation for beer. All that would demonstrate is that Matthew's subjectivity has integrated or developed a unity with his new body. The limitation with the third-person point of view, which detaches the body from the subject, is that such a view over-abstracts our subjectivity (however it is constituted). Rather, I suggest that our experiences, and our choices, goals, plans, intentions, and desires are *lived and embodied* experiences, choices, goals, plans, intentions, and desires.

The overall contention here is that our rights of use and control of our body enable preferences and choices that cannot exist independently of us as the self-owner. Moreover, the necessary relationship between the self-owner and such preferences and choices is not merely attributable to the close and immediate (and perhaps temporary) connection between a person's subjectivity

[77] M Merleau-Ponty, *Phenomenology of Perception* (n 74) 167.

and their body. Rather, the necessary relationship between the self-owner and the preferences and choices that the rights of self-ownership enable is attributable to our physical subjectivity. Our subjectivity—the features and qualities about us that ground moral duties towards us—are features and qualities that are interwoven with our bodies. Recall that, for Merleau-Ponty, *being* and *owning* are distinct categories (see pages 61–2). Either something stands in external relation to our bodies (owning) or something is already interwoven in a task of engaging with the world (being). As Taylor Carman explains with regards to Merleau-Ponty's view:

> [T]he body is not a kind of quasi-object thing with which I identify thanks to the localisation in it of my subjective sensations, rather the attribution of sensations to myself in the first place presupposes my own prior identification with my body.[78]

The body, unlike other objects, is not something that we experience, and is not merely experienced as closely and immediately connected to us. Rather, prior to being experienced, the body is *the way in which* we experience things. The body, in this way, is necessarily associated with the subject and the preferences and choices that the use and control of our bodies enables are preferences and choices that are necessarily associated with us.

We can appreciate how property rights and the right to bodily integrity fundamentally differ on a conceptual basis. In terms of particular legal actions, property rights-based actions (such as trespass to goods) protect rights that are contingent to the rights-holder. Torts against the person (such as trespass to the person) protect rights that are necessarily associated with the rights-holder. As we may be able to anticipate, this conceptual distinction is able to explain how the legal structure of property rights differs from the structure of the right to bodily integrity. Given the distinction between preferences and choices that can or cannot exist independently of the rights-holder, the question that we will be concerned with in latter chapters is whether rights in bodily material are contingent or non-contingent rights.

c. Contingency and Value Equivalence

In forthcoming chapters we will encounter the temptation to dissolve or collapse this conceptual distinction between contingent and non-contingent rights. The temptation follows from the practical ease with which property

[78] T Carman, 'The Body in Husserl and Merleau-Ponty' (1999) 27(2) *Phil Topics* 205, 222.

rights fill the void in the law left by the 'no property rule'. Allow me to briefly fortify the distinction between contingent and non-contingent rights by mapping this conceptual distinction onto the distinction between things that have value equivalence and things that have value non-equivalence.

Recall that a value is denigrated where 'the relevant goods cannot be aligned along a single metric without doing violence to our considered judgment about how these goods are best characterised'.[79] In a similar way to denigrating the value of something that has no value-equivalent, collapsing the distinction between contingent and non-contingent rights undermines the preferences and choices that are particular to the rights-holder. When we consider Joanna's ability to possess and control her blood test results, or her ability to control others contact with, and intrusions into, her body, we encounter a set of preferences and choices that have value in their own right. Lucy cannot stand in the same position as Joanna with regards to the control of these things because the control of these things protects a value that is non-equivalent to whatever value Lucy could obtain from controlling or using Joanna's personal information or using or interfering with her body. To view these non-contingent rights as contingent rights, equates to treating these relationships of control that have value in their own right as having an equivalence with something else.

The fundamental concern here is that our relationship with our bodies 'constitutes the condition under which alone something can be an end in itself and has not merely a relative worth'.[80] Where our relationship with our body give rises to legally enforceable rights, such rights are necessarily associated with the rights-holder, and are therefore non-contingent rights. If we were to dissolve or collapse the distinction between contingent and non-contingent rights, we would be reducing all of our connections or intentional ties with the world into a set of relationships with things that can be 'aligned along a single metric'.[81] Such a reduction does violence to 'our considered judgment about how [some] goods are best characterised'.[82]

[79] CR Sunstein, 'Incommensurability and Kinds of Valuation: Some Applications in Law' in R Chang (ed) *Incommensurability, Incomparability and Practical Reason* (Harvard University Press, Cambridge 1997) 234, 238.

[80] MJ Gregor (ed), I Kant, *The Groundwork of the Metaphysics of Morals* (Cambridge University Press, Cambridge 1996) 42–3.

[81] CR Sunstein, 'Incommensurability and Kinds of Valuation' (n 79) 238.

[82] CR Sunstein, 'Incommensurability and Kinds of Valuation' (n 79) 238.

V. Chapter Summary

This chapter has identified two conceptual features of property rights and then considered how these features align with the right to bodily integrity. I have argued that property rights are exclusive rights. Three premises underlie this explanation. It is because we have an interest in the purposeful use of objects or resources that the law formulates a legal relationship (between rights-holder and duty-bearer) that has the object or resource as its focal point. Rather than cataloguing the various ways in which the rights-holder can use the object or resource, the interest in the object or resource is taken to be an interest in an open-set of activities. The law then protects this sphere of undefined activity through the exclusion of others from the object or resource. Hence property rights operate through the construction of an exclusionary boundary around the object or resource.

The same is true for our right to bodily integrity. We exercise our autonomy, promote our welfare, or maximize our preference satisfaction through purposely using our bodies in particular ways in pursuit our plans, preferences, goals, and interests. The various ways in which we use our bodies, in pursuit of various plans, preferences, goals, and interests cannot be exhaustively listed. The use of our body is therefore treated as a sphere of undefined activity that is protected through the exclusion of others. Hence, our right to bodily integrity represents our ultimate and exclusive agenda-setting authority over the use and control of our bodies.

The second conceptual feature described property rights as contingent rights. Property rights are contingent rights since they protect preferences and choices that have no particular relationship with the rights-holder. There are two criteria that can be used to isolate this distinction. Where there are no normative consequences (beyond mere allocative consequences) following the change in rights-holder then such rights will be contingent rights. This is because, for contingent rights, the subsequent rights-holder is able to 'stand in the same position' with regards to the object or resource as the original rights-holder. Our property rights are rights that can exist independently of us. In comparison, our rights to bodily integrity, privacy, and reputation are rights that are necessarily associated with us.

The explanation for why our right to bodily integrity is a right that is necessarily associated with us is important. I have considered two possible explanations. The necessary association can be explained in terms of the close and immediate connection between a person's subjectivity and their body. However, this explanation only satisfies the first, and not the second, criteria of contingency. Although (at any one point in time) another person cannot stand in the same position as the self-owner with regards to the self-owner's body, there would only be mere allocative consequences if the body (or a body) were to be transferred to another. I have suggested that this view of the body over-abstracts a person's subjectivity and thereby drains the body of its value. From a first-person perspective, the necessary association between the right to bodily integrity and the rights-holder follows from the unity or integration of the subject with his or her physical and material environment. I have also sought to emphasize the importance of the conceptual distinction between contingent and non-contingent rights; non-contingent rights protect a relationship with a thing that has value in its own right, and such relationships ought not be treated as something that has a value equivalence with any other ownership relationship.

The recognition of property rights in semen provided the Supreme Court of New South Wales in *Re: Edwards* two manoeuvres in legal reasoning that were unavailable to the Court of Appeal of England and Wales in *ex parte Blood*.[83] First, the recognition of property rights introduced a set of rights beyond those stipulated in the legislation. Rather than legislation determining whether gametes could be stored, transferred, or used, the statutory regime in *Re: Edwards* was held to exist alongside the exclusion strategy of property law. The exclusion strategy provided a right for Mrs Edwards to possess Mr Edwards's bodily material beyond the rights stipulated in governing legislation. Second, the court's recognition of property rights provided a means of transferring the right to possess the semen to the widow of the progenitor. According to Munzer, 'property rights are rights that protect the choice to transfer'.[84] This structural feature reflects a conceptual assumption that property rights are rights that are contingent to the rights-holder.

Finally, note how this conceptual assumption is misplaced; the semen in *Re: Edwards* enables a set of preferences and choices that can only be exercised by the progenitor (Mr Edwards) and a co-participant in assisted reproductive therapy (Ms Edwards). This conceptual inconsistency between the rights that arise in the bodily material and property rights has structural and doctrinal implications, and this conceptual (and structural) inconsistency provides the main impetus for resisting the wholesale application of property law to the use and storage of bodily material.

[83] *R v Human Fertilisation and Embryology Authority, ex parte Blood* [1997] 2 All ER 687; *Jocelyn Edwards; Re the Estate of the Late Mark Edwards* [2011] NSWSC 478.

[84] SE Munzer, *A Theory of Property* (n 51) 48.

5

The Structure of Property Law

Mr Kelly was an artist who was permitted to draw anatomical specimens held by the Royal College of Surgeons. The specimens were held by the College for use by doctors training to be surgeons. Kelly removed 35 to 40 body parts from the College. He made casts of the body parts and then either buried them in a field, stored them in his attic, or stored them in his friend's basement. Kelly was charged with theft.

Mr Kelly argued in his defence that the body parts could not be stolen. This claim was made on the basis that the body parts were not items of property and that the College did not have lawful possession of them. The Court of Appeal of England and Wales recognized an exception to the 'no property rule': that parts of corpses are capable of being property if they had acquired different attributes by virtue of the application of skill.[1] The Court of Appeal was then able to find that the body parts were items of property within the meaning of s 4(1) of the Theft Act 1968,[2] and that the College had sufficient possession for the purposes of s 5(1) of the Theft Act 1968.[3]

The College therefore had the right to possess the anatomical specimens. The application of the 'work or skill' exception to the 'no property rule' meant that the College possessed the body parts *as items of property*. The effect of the body parts obtaining the status of property was to impose a duty on all other persons not to 'dishonestly appropriate' the body parts and for the dishonest appropriation to be itself a legal wrong that Kelly could be prosecuted for.[4] As items of property, the specimens were protected by a particular legal structure. Here we will identify the legal structure of property law.

[1] *R v Kelly/R v Lindsay* [1998] 3 All ER 741, 749.

[2] Theft Act 1968, s 4(1): '"Property" includes money and all other property, real or personal, including things in action and other intangible property.'

[3] Theft Act 1968, s 5(1): 'Property shall be regarded as belonging to any person having possession or control of it, or having in it any proprietary right or interest ...'

[4] Theft Act 1968, s 1(1): 'A person is guilty of theft if he dishonestly appropriates property belonging to another with the intention of permanently depriving the other of it ...'

The purpose of this chapter is threefold: to identify the 'structural' or 'doctrinal' features of property law, compare these structural features with other branches of law, and consider how these structural features have so far been applied to the use and storage of bodily material. As I will explain, the structural features of property law follow from the conceptual features of property law. Recall that property rights have two conceptual features: property rights exclude all others from an object or resource and protect preferences and choices that can exist independently of the rights-holder. The conceptualisation of property as a right to exclude others from the object or resource is able to explain why, for property rights:

the basis of the right is the relationship with the thing itself rather than the relationship with the duty-bearer,

the content of the right is the relationship with the thing itself rather the value that can be derived from the thing, and

the content of the primary duty is a duty of non-interference rather than a duty of care.

In addition, it is because property rights can be conceptualized as protecting preferences and choices that can exist independently of the rights-holder, we can understand why, for property rights:

the content of the remedial duty is to correct the misallocation of entitlements rather than distribute benefits and burdens on upon a criterion of merit, and

the transferability of the right is permissible (or permissible under certain circumstances) rather than the right being inalienable.

These structural features will help explain why there is pressure on the common law to recognize property rights in bodily material. Simply put, entitlement-holders benefit from the exclusionary boundary that property law is able to provide. Yet, property law is not the only branch of law that employs an exclusion boundary and there may be good reasons why we would want to avoid the wholesale application of property law to the use and storage of bodily material. The reasons for resisting the application of property law to the use and storage of bodily material can be initially identified in this chapter. Such reasons concern both the structure of the basic inequality between the rights-holder and the duty-bearer that follows from the exclusion strategy and the structural features that follow from orientation of property rights as protecting preferences and choices that can exist independently of the rights-holder.

Note that some property rights-based actions, such as conversion and trespass to goods, protect entitlements in things through these three structural features of the exclusionary boundary. Other actions, such as negligent damage to property and bailment, are based upon an overlap of the exclusion and governance strategies. As a result of this overlap, these actions represent a structural hybrid. The structural distinctions that follow from the contingency or non-contingency of rights also represent a series of complex distinctions. There is a great degree of subtlety and variation between the various property law doctrines. Without overlooking these subtleties, the focus here is developing the main dichotomy of benefiting from the legal structure of the exclusion strategy without having to also apply a legal structure that presumes that the entitlements in the object or resource represents preferences and choices that can exist independently of the rights-holder.

I. The Basis of the Right

In the previous chapter I explored a distinction between the exclusion strategy and the governance strategy. According to this conceptual distinction, the exclusion strategy excludes all others, except the entitlement-holders themselves, from the object or resource. In comparison, the governance strategy allocates rights and duties between participants in an activity or between the parties to an interaction. As I will explain in this section, it is because property law focuses on the exclusion of others from the thing, the right is exercisable against an open set of persons. For some this characteristic, that a property right 'is enforceable against third parties', is 'the defining characteristic of a property interest'.[5]

a. The Exclusion Strategy and Pre-Existing Rights

The structure of property rights includes 'trespassory protection'[6] or 'tresspassory rules'[7] which operate by imposing 'on anyone other than the owner a duty not to interfere without the owner's consent'.[8] This structure of protection can be contrasted with 'transactional protection', which is 'a facet of a relationship established by the transaction creating the right in question'.[9] This distinction can also be understood as between 'direct rights' which are rights that arise 'against [the duty-bearer] because of [the duty-bearer's]

[5] J Hill, 'The Proprietary Character of Possession' in E Cooke (ed), *Modern Studies in Property Law: Volume 1* (Hart Publishing, Portland 2001) 22.

[6] JE Penner, *The Idea of Property in Law* (Oxford University Press, Oxford 1997) 128.

[7] JW Harris, *Property and Justice* (Clarendon Press, Oxford 1996) 5.

[8] JE Penner, *The Idea of Property in Law* (n 6) 128.

[9] JE Penner, *The Idea of Property in Law* (n 6) 128.

own conduct', and 'pre-existing rights', which are rights that are enforceable against the duty-bearer independent of the duty-bearer's conduct.[10]

Despite various formulations of the distinction, there is convergence as to 'the *effect* of a property right':[11] that the right is (prima facie)[12] enforceable against an open set of persons since the success of the action 'does not depend on there being a [prior] relationship between the holder of the right and the person infringing it'.[13] Hence, for property rights-based actions, the legal right arises because the right-holder's relationship with the thing is the *source or basis* of the duty imposed on the duty-bearer. This right can be described as 'pre-existing' because the right-holder's interest in the thing existed prior to the interaction with the duty-bearer. This structural feature of property law follows from the exclusion strategy that is employed by property law. Since exclusive rights represent a sphere of 'indefinite and undefined activity',[14] the correlative duties are not derived from, or determined by, any particular activity related to the object or resource.

Actions for conversion, trespass to goods, reversionary injury, theft, and trespass to the person all concern pre-existing rights. The effect of this is to impose a duty on an open set of persons. As mentioned above, since the body parts acquired by the defendant in *R v Kelly* were deemed to be items of property, s 1 of the Theft Act 1968 prohibited the defendant from 'dishonestly appropriating' the body parts. [15] It did not matter who the defendant in *R v Kelly* was, the circumstances of the transaction, nor his or her relationship with the Royal College of Surgeons. The duty was imposed on Mr Kelly as the corollary of recognizing the College's proprietary right in the body parts. Similarly, under an action for conversion, provided the claimant has the immediate right to possession,[16] an open set of persons are under a duty

[10] B McFarlane, *The Structure of Property Law* (Hart Publishing, Oxford 2008) 20–2.

[11] B McFarlane, *The Structure of Property Law* (n 10) 139 [emphasis added].

[12] B McFarlane, *The Structure of Property Law* (n 10) 179: As McFarlane explains, the right is *prima facie* enforceable given the possibility that the defendant has a defence to a property rights claim.

[13] JE Penner, *The Idea of Property in Law* (n 6) 128: '…does not depend on there being a *pre-existing* relationship between the holder of the right and the person infringing it.'

[14] JE Penner, *The Idea of Property in Law* (n 6) 72.

[15] Theft Act 1968, s 1(1): 'A person is guilty of theft if he dishonestly appropriates property belonging to another with the intention of permanently depriving the other of it…'

[16] S Green, 'Understanding the Wrongful Interference Actions' (2010) 74 *Conv* 15, 19–20: '[T]respass and conversion…are available only to the individual with the immediate possessory right in an asset, and not to anyone with a superior proprietary, but lesser possessory, interest.'

not to deprive the rights-holder the full benefit of the right.[17] Under trespass to goods, where the claimant has actual possession,[18] an open set of persons are under a duty not to directly interfere with the goods.

The actions that protect the right to bodily integrity also share this structural feature with these property rights-based actions. The duty not to threaten the application of force to the body, the duty to avoid intentional and non-consensual contact, and the duty not to interfere with the free movement of the body are all pre-existing duties. That is, they are duties imposed on an open set of persons independent of any interaction between the rights-holder and the duty-bearer.

b. The Governance Strategy and Direct Rights

The governance strategy employs a different structural approach. Under the governance strategy, the duty on a duty-bearer arises because of a particular relationship, transaction, or interaction between the rights-holder and the duty-bearer. This right can be described as arising 'directly' against the duty-bearer because of the relationship between the two actors.[19] Since the governance strategy is concerned with the apportionment of rights and duties between parties to an activity, transaction, or interaction, the duty imposed on the duty-bearer arises 'directly' out of the activity, transaction, or interaction.

Negligence, contract, fiduciary duties, and traditional doctrines of confidentiality formulate direct rights that are enforceable against particular duty-bearers. Even where there is a recognized property right, an action for negligent interference with the item of property requires a series of factual findings that concern the relationship between the rights-holder and the duty-bearer. Similarly, actions for breach of confidence, or negligence causing personal injury also require a particular transaction, relationship, or interaction to form the basis of a duty.

For example, in *Moore v Regents of the University of California*, the patient's rights of control over his body ought to have been facilitated by the duties owed to him by virtue of the relationship between him as the patient and the defendant (as his physician).[20] In comparison, the donors of bodily material

[17] S Green, 'Understanding the Wrongful Interference Actions' (n 16) 20: '[C]onversion is available not just against anyone who still has possession of the converted asset, but against anyone who ever has interfered with a claimant's possessory rights over it.'

[18] S Green, 'Understanding the Wrongful Interference Actions' (n 16) 19–20.

[19] B McFarlane, *The Structure of Property Law* (n 10) 20–2.

[20] *Moore v Regents of the University of California* 793 P 2d 478 (Cal 1990) 128–32.

in *Greenberg v Miami Children's Hospital* were unable to establish that the subsequent possessors of their bodily material owed them a duty. This is because the donors were unable to establish that a fiduciary duty applied to the researchers of their condition disease, the court finding that '[t]here is no automatic fiduciary relationship that attaches when a researcher accepts medical donations and the acceptance of trust'.[21]

In *Yearworth v North Bristol NHS*, both the successful action for negligent damage to property and the unsuccessful action for negligence causing distress were dependent upon the NHS Trust owing a duty of care to the claimants.[22] Similarly, in *Re: Organ Retention* the claimants had to establish that a duty of care arose in the context of hospital post-mortems. More precisely, the claimants had to establish that a doctor could owe a duty of care to the mother after the death of her child with regards to the post-mortem examination of the child's body.[23]

An action in bailment, like in negligence and traditional duties of confidentiality, only arises where there is a particular interaction or transaction between the rights-holder and the duty-bearer. Hence, the Court of Appeal in *Yearworth v North Bristol NHS* also held that obligations on the defendant arose because they took possession of the semen in circumstances that 'involve[d] an assumption of responsibility of the safe keeping of the goods'.[24] Hence, the finding of a bailment relationship also requires a factual inquiry into the circumstances of the transfer of possession that took place between the claimant and defendant.

We can therefore map a structural distinction between pre-existing rights and direct rights as it applies to the use and storage of bodily material. A pre-existing right is exercisable against an open set of persons. This explains the employment of the common law 'work or skill' exception to the 'no property rule'. In both *Doodeward v Spence* and *R v Kelly*, the 'work or skill' exception was applied so as to ground a right of possession that was enforceable against the defendants independent of any particular transaction

[21] *Greenberg v Miami Children's Hospital* 264 F Supp 2d 1017.

[22] *Yearworth v North Bristol NHS Trust* [2009] EWCA Civ 37 [13].

[23] *Re: Organ Retention Group Litigation* [2005] QB 506, [170]–[206], [206]: 'Once the doctor-patient relationship is established, as I hold it is, in my view, the clinician owed a duty of care when seeking consent for a post-mortem examination … In the circumstances, I hold that the duty of care extended to giving the parents an explanation of the purpose of the post-mortem and what it involved including alerting them to the fact that organs might be retained.'

[24] *Yearworth v North Bristol NHS* (n 22) [48(c)]: 'Thus "the obligation arises because the taking of possession in the circumstances involves an assumption of responsibility for the safe keeping of the goods": see the advice of the Privy Council in *Gilchrist Watt and Sanderson Pty Ltd v York Products Pty Ltd* [1970] 1 WLR 1262 at 1268H.'

or interaction. It is this particular structural feature of property rights that has motivated the first main exception to the no property rule. As we have seen with reference to *Moore v Regents of the University of California, Greenberg v Miami Children's Hospital, Washington University v Catalona,* and *Re: Organ Retention,* in the absence of a pre-existing right, the claimant is required to establish that a direct right arises on the basis of the relationship between the claimant and defendant.

II. The Content of the Right

We know from the previous chapter that the exclusion strategy protects an undefined sphere of activity whereas the governance strategy protects particular activities or interactions. If I have a right to possess a thing, different branches of law will protect different dimensions of my right of possession; the law may protect my possession per se or the law may protect my right to possess the thing in the context of a particular activity or interaction. In other words, my right to the thing may have a different *content* depending on the applicable branch of law.

a. The Original and Derivative Dimensions of the Right

There is a structural distinction between branches of law that protect the 'original' dimension of a right and branches of law that protect only the 'derivative' dimension of a right.[25] This distinction can also be understood in terms of actions that are 'actionable *per se*' and actions that are 'actionable upon proof damage'.[26] However, here the language of 'damages' is reserved for a separate inquiry as to the remedial duties. As Donal Nolan explains, 'the issue of the damage sufficient to establish a cause of action should not be confused with the harms for which recovery is permitted once the cause of action has been established'.[27] Here we are concerned with the damage that is sufficient to establish a cause of action. We consider at pages 157–65 the harms or losses for which recovery is permitted.

[25] S Green, 'Rights and Wrongs: An Introduction to the Wrongful Interference Actions' in D Nolan and A Robertson (eds), *Rights and Private Law* (Hart Publishing, Oxford 2012) 538.

[26] D Nolan, 'New Forms of Damage in Negligence' (2007) 70 MLR 59, 60: 'The first is the rather obvious observation that some torts are actionable per se and some are actionable only on proof of damage.'

[27] D Nolan, 'New Forms of Damage in Negligence' (n 26) 61.

The 'original' dimension of a right or entitlement concerns the relationship between the rights-holder and the thing. For instance, Matthew may have the right to physical control, or the right to obtain physical control, over a painting. Mark may have the right to determine how, and by whom, the painting is used. Lucy may have the right to transfer all of these rights to the painting to another person. Since these rights concern the relationship with the thing (the painting), they are 'right[s] *to* the subject matter'.[28] When someone is deprived of the 'right *to* a thing' they suffer—generally speaking—a loss. The loss they suffer, however, is intangible. Depriving Matthew of his rightful possession of the painting, disregarding Mark's valid instructions as to the use of the painting, or preventing Lucy from transferring the painting to another person, are all instances of interference with the relationship between the rights-holder and the thing. Yet, it is difficult to quantify the harm done, or loss suffered, that is caused by merely disrupting the relationship between Matthew, Mark, or Lucy, and the painting.

In such instances of interference, Matthew, Mark, and Lucy will often be able to identify a further loss that follows from the interference with the 'right *to* the painting'; it is because Matthew was deprived of possession of the painting that Matthew could not participate in an exhibition, it is because (contrary to Mark's instructions) the painting was exposed to regular sunlight that the painting is discoloured and has depreciated in value, it is because Lucy could not transfer the painting overseas that Lucy could not benefit from a lucrative contract. These more tangible losses are the 'derivative' dimensions of the right.[29] The tangible value of having entitlements consists of having derivative entitlements in the object, resource, or asset; the ability to display a painting, to preserve a painting, or to profit from the sale of a painting. In short, an interest *in* the subject matter.[30]

b. Exclusion, Governance, and the Content of the Right

Protection of both the original and derivative dimensions of the right follows from the exclusion strategy of property law. Since exclusive rights concern the exercise of open-ended entitlements, interference with the relationship between the person and thing is *itself* a wrongful intrusion into the sphere of protected activity. Depriving owners of their rights of possession, and acting inconsistently with their instructions as to the use or care of the object, are actionable wrongs. For exclusive rights, the law performs a 'qualitative'

[28] S Green, 'Rights and Wrongs' (n 25) 538.
[29] S Green, 'Rights and Wrongs' (n 25) 538.
[30] S Green, 'Rights and Wrongs' (n 25) 538.

assessment to determine that 'either an individual has had a right infringed, or he has not'.[31] If the right has been infringed, then the individual can recover for the loss in original and derivative entitlements.

In contrast, where the law employs the governance strategy the focus of the law is on the particular activity of participating in an exhibition, preserving a painting, or selling a painting; the relationship between rights-holder and thing is not the focal point of the law's attention. Rather, the law here is concerned with particular actions, transactions, or activities. Only the interests in such activities are legally protected. Here the law is protecting particular activities or interaction and performs a 'quantitative assessment' of whether the claimant's interest in the activity or interaction was interfered with by the defendant.[32]

Conversion, reversionary injury, trespass to goods, and trespass to the person all concern original rights. We can identify the content of the right by looking to the damages that are sufficient to establish a cause of action. Recoverable losses include both loss from the interference with the good and losses consequential to (or derived from) the interference with the good.[33] The key observation that Sarah Green provides is that consequential loss 'whilst potentially recoverable, is not necessary for the action to succeed, since this is not its primary concern'.[34] Rather, the primary concern of personal property law is the relationship with the thing. Wrongful interference with goods is therefore, by definition, concerned with 'damage *to* goods or to an interest *in* goods'.[35] In other words, to succeed in a claim in conversion, trespass, or reversionary injury, it is sufficient to show that the loss suffered concerned interference with the possession or control relationship with the thing. As Robert Stevens explains, 'There need be no loss *consequent* upon the defendant's interference'.[36]

The right to bodily integrity also protects original rights. To succeed in a claim for assault, battery, or unlawful detention, it is sufficient to show that there was the threat or the application of force to the body or the deprivation of the free movement of the body. The interference with the right to bodily integrity is itself sufficient for an action, without having to show a loss derived from the interference. As Jonathan Herring explains in the medical context, unlike a claim in negligence causing personal injury, 'in a case of

[31] S Green, 'Rights and Wrongs' (n 25) 536.
[32] S Green, 'Rights and Wrongs' (n 25) 536.
[33] Torts (Interference with Goods) Act 1977, s 3(2).
[34] S Green, 'Understanding the Wrongful Interference Actions' (n 16) 21.
[35] Torts (Interference with Goods) Act 1977, s 1(d) [emphasis added].
[36] R Stevens, *Torts and Rights* (Oxford University Press, Oxford 2007) 66.

battery there is no need to show that a patient suffered loss because the battery will be itself a legal wrong'.[37]

Although negligence actions have so far been characterized as employing the governance strategy, negligent damage *to property* nonetheless protects original rights. This is because claims in negligent damage to property have 'dual sources of rights and duties'.[38] Negligence provides a *basis* for a defendant's duty not to interfere with an existing asset (a 'direct-duty'), whilst the 'existence and content' of the correlative right 'is determined by the law of property'.[39] This means that the *content of the right* concerns both the original and derivative dimensions of the entitlement. Or, as Stevens explains, 'damages for negligent damage to property are concerned with the vindication of rights, and not with compensation for loss'.[40] An action for negligent damage to goods therefore represents an overlap between the recognition of existing property rights and the right to recover for losses that are caused by wrongful conduct. 'The problem' that Green identifies, 'is that the original right is alien to the law of torts as it is native to the law of property'.[41] Whereas property rights-based actions recognize the deprivation of original rights, the remaining torts are generally unwilling to recognize the intangible loss that is suffered from (merely) the interference with the relationship between rights-holder and thing. This is because, Green explains, 'the law relating to wrongs is more comfortable with having something tangible to remedy, than it is with protecting the very existence of relationships'.[42]

c. The Content of the Right and Bodily Material

Unlike property-based actions or trespass to the person, 'to succeed in a claim in negligence' in the medical context 'it must be shown that the patient suffered some harm'.[43] The reluctance of non-property actions to identify loss in the form of deprivation of 'original entitlements' means that claimants, who are unable to bring a property rights-based action, must prove a

[37] J Herring, *Medical Law and Ethics* (4th edn, Oxford University Press, Oxford 2011) 152.

[38] S Green, 'Rights and Wrongs' (n 25) 541.

[39] S Green, 'Understanding the Wrongful Interference Actions' (n 16) 16: 'Once the action has been triggered by a breach of the relevant duty, the process then involves a consideration of the content of the claimant's right.'

[40] R Stevens, *Torts and Rights* (n 36) 73. See also D Nolan, 'New Forms of Damage in Negligence' (n 26) 61: 'The fourth point is that physical harm to the claimant's person or property clearly constitutes actionable damage for negligence purposes, though the boundaries of the concept of physical damage are not always clear.'

[41] S Green, 'Rights and Wrongs' (n 25) 538.

[42] S Green, 'Rights and Wrongs' (n 25) 538.

[43] J Herring, *Medical Law and Ethics* (n 37) 152.

'derivative loss' in order to succeed in a non-property action. The two negligence claims in *Yearworth v North Bristol NHS* are illustrative of the difference in approaches to actionable loss. The patients' claim for negligence causing injury or distress failed because they could not prove that, on the balance of probabilities, the damage to the semen interfered with their ability to conceive a child. According to the Court of Appeal:

> [T]he decision in *Gregg v. Scott* . . . compelled a conclusion that, unless the men were to demonstrate a greater than even chance that the lost sperm could have been used in order to achieve conception, then, irrespective of any recovery of their natural fertility, they could not claim in respect of such physical damage to their overall ability to become fathers.[44]

In other words, the claimants were unable to prove a harm or loss in a derivative sense. Once the semen in *Yearworth v North Bristol NHS* was conceived of as items of property, the claimants were able to succeed in their negligence claim since they were able to prove a loss in terms of an interference with their 'legal ownership or possessory title'.[45] That is, interference with an 'original' right.

A distinction therefore exists between property-based actions, which recognize loss in the deprivation of an original entitlement, and loss-based actions, which require proof of loss in a derivative entitlement. The willingness of property law to identify an actionable loss in terms of disrupting the relationship between the rights-holder and the item of property contributes to the stringency of protection provided by the exclusionary boundary. For Remigius Nwabueze, it is this feature of the common law that makes the use of property rules for regulating the donation of organs necessary if 'only to avoid the remedial hurdle of proving a contemporaneous ... loss in negligence'.[46] Nwabueze correctly identifies that in instances where an organ for donation is misdirected, misapplied, or maliciously destroyed, it is difficult to find an actionable loss under non-property based rules. His quasi-proprietary solution is to avoid the remedial hurdle of 'proving the existence of a physical or pecuniary injury necessary to ground an action in negligence ... by attributing to the claimant a quasi-property right' in the bodily material.[47]

[44] *Yearworth v North Bristol NHS* (n 22) [24].

[45] *Yearworth v North Bristol NHS* (n 22) [25]: 'In order to enable a person to claim in negligence for loss caused to him by reason of loss of or damage to property, he must have had either the legal ownership of or a possessory title to the property concerned at the time when the loss or damage occurred.' Per Lord Brandon in *Leigh and Sillavan Ltd v Aliakmon Shipping Co Ltd* [1986] AC 785 at 809F.

[46] RN Nwabueze, 'Donated Organs Property Rights and the Remedial Quagmire' (2008) 16 *Med L Rev* 201, 205.

[47] RN Nwabueze, 'Remedial Quagmire' (n 46) 205.

Our second structural distinction is therefore concerned with the content of the right. The exclusion strategy protects rights that concern the relationship between the person and the thing, as well as any benefits derived from the relationship. It follows that the interference with items of property, or the infringement of the right to bodily integrity, is itself sufficient to prove an actionable loss. It is this feature of property rights that motivated the recognition of property rights in *Yearworth v North Bristol NHS*. Since the patients were unable to show that the inadequate storage of the semen caused a consequential loss (opportunity to conceive), the success of the claim depended upon establishing an actionable loss by showing that inadequate storage deprived the patients of their 'legal ownership or possessory title' under property law.[48] By convincing the court to treat their semen as an item of property, the claimants were able to show that the inadequate storage of their semen caused a 'physical deprivation'.[49] It was ultimately this physical deprivation—or interference with an 'original right'—that enabled the claimants to succeed in their claim against the NHS Trust. In contrast, the rights and duties under the governance strategy only concern the activities derived from the use of a thing. Under the governance strategy, proving an actionable loss creates a 'remedial hurdle' for the claimant.

III. The Content of Primary Duties

The third structural feature that follows from the exclusion strategy of property law concerns the content of the duty that is imposed on the duty-bearer. Where a right arises, a correlative duty will also arise in order to protect the right. Yet, what constitutes conduct that is inconsistent with the right varies between branches of private law. In other words, legal actions differ in terms of the type of wrongful actions by a defendant that are sufficient for a successful claim against a defendant.

a. Duties of Non-Interference and Duties of Care

The distinction drawn here is between duties of non-interference and duties of care. Duties of non-interference impose a duty to abstain from voluntary action that interferes with the rights of the rights-holder.[50] To adhere to such duties requires inaction or the abstinence from a particular set of actions.

[48] *Yearworth v North Bristol NHS* (n 22) [25].
[49] RN Nwabueze, 'Remedial Quagmire' (n 46) 206.
[50] See P Cane, 'Causing Conversion' (2002) 118 *LQR* 544.

Duties of non-interference are therefore concerned with the outcome of the duty-bearers voluntary or deliberate conduct. In other words, such duties impose an 'obligation to succeed' [51] in acting in a way that is consistent with the entitlements of the rights-holder. In contrast, where a duty-bearer is under a duty of care, the duty-bearer is under an obligation to take action that is sufficient to safeguard the rights-holder's interests. Such duties are concerned with the combination of the outcome of the duty-bearer's conduct (their 'obligation to succeed') and the standard of conduct undertaken by the duty-bearer (their 'obligation to try').[52]

An action for trespass to goods requires direct and wrongful interference with the claimant's goods. An action for conversion requires more; the interference must amount to the appropriation of the full benefit of the claimant's entitlements.[53] Although the duties imposed under trespass and conversion differ in terms of the degree or extent of interference, the content of these duties both require inaction or abstinence. That is to say the only way of performing the duty under these property rights-based actions is to not undertake voluntary conduct that directly interferes with, or appropriates, the goods. Note that although Lord Nichols in *Kuwait Airways* describes conversion in terms of 'deliberate' conduct,[54] the conduct need only be deliberate in 'the sense that the [interferer] deliberately performed the act'[55] and not in the sense that the interferer deliberately or intentionally infringed the rights of the rights-holder.[56]

Duties of non-interference are also imposed under trespass to the person. A battery is committed where the defendant intentionally causes physical

[51] J Gardner, 'Obligations and Outcomes in the Law of Torts' in P Cane and J Gardner (eds), *Relating to Responsibility: Essays for Tony Honoré on His Eightieth Birthday* (Hart Publishing, Oxford 2001) 10: 'An obligation to succeed is a straightforward obligation to φ, for the performance of which only one's actually φing matters, never mind what steps one takes (if any) with a view to φing.

[52] J Gardner, 'Obligations and Outcomes in the Law of Torts' (n 51) 10: 'An obligation to try is the converse case: only the steps one takes with a view to φing are relevant to whether one performs the obligation, never mind whether one actually φs thereby.'

[53] S Green, 'Rights and Wrongs' (n 25) 528–9.

[54] See *Kuwait Airways Corp v Iraqi Airways Co (No 6)* [2002] UKHL 19, [2002] 2 AC 883 (Lord Nicholls) [39]: 'Conversion of goods can occur in so many different circumstances that framing a precise definition of universal application is well nigh impossible. In general, the basic features of the tort are threefold. First, the defendant's conduct was inconsistent with the rights of the owner (or other person entitled to possession). Second, the conduct was deliberate (not accidental). Third, the conduct was so extensive an encroachment on the rights of the owner as to exclude him from use and possession of the goods.'

[55] S Green, 'Rights and Wrongs' (n 25) 532.

[56] P Cane, 'Causing Conversion' (n 50): 'The fundamental point is that conversion consists of conduct inconsistent with the owner's rights regardless of the intention with which it is done.'

contact.[57] The intention required concerns only the deliberateness of the conduct, and not an intention to cause the harm that follows from non-consensual contact. The duty imposed under battery is therefore a requirement of inaction; not to undertake an action that causes non-consensual physical contact. It is enough to show that the defendant did something wrongful in the form of deliberate contact with the person.

The requirement of intentional interference for a breach of a duty of non-interference can explain why the duty of non-interference is a strict duty. The act of deliberately interfering with, or appropriating, the goods, or the act of deliberate physical contact with another, is an intrusive wrong even in the absence of fault. As Simon Douglas explains with regards to conversion, strict duties of non-interference account for a particular set of intentional (deliberate) intrusions without representing 'too great a restriction on the defendant's freedom of action' since it is only when a 'defendant intentionally interferes with a chattel that he can be liable in conversion. Unintended interferences must be litigated in negligence where fault must be proved'.[58]

Hence, another set of duties, which are complementary to duties of non-interference, arise to protect property rights and the right to bodily integrity that require a finding of fault. For instance, the content of the duties imposed under negligence concerns the standard of care taken by the duty-bearer and resulting interference with the rights of the rights-holder. To succeed in their claim in bailment and for negligence damage to property, the claimants in *Yearworth v North Bristol NHS* were required to show that as the NHS Trust failed to take reasonable steps to safeguard their proprietary or possessory interest in the stored semen.[59] Similarly, the claimants

[57] FA Trindade, 'Intentional Torts: Some Thoughts on Assault and Battery' (1982) 2 *Oxford J Legal Stud* 211, 220: 'In battery what is required is intentional contact not an intention to do harm—and it is not correct to say that trespass can only be brought for the direct physical infliction of harm.'

[58] S Douglas, 'The Nature of Conversion' (2009) 68 *CLJ* 198, 220: 'The importance of a claimant's property rights may still justify the imposition of strict liability in the tort but only if this does not represent too great a restriction on the defendant's freedom of action. It is suggested that it does not. It does not because it is only when a defendant intentionally interferes with a chattel that he can be liable in conversion. Unintended interferences must be litigated in negligence where fault must be proved. As a result a defendant's general freedom to act is still recognised *except* when [the defendant] chooses to consciously interfere with a chattel. Because this is a well defined and limited form of activity, his freedom is restricted in a comparatively minor way. The outcome is that a balance is struck between the importance attached to the claimant's property rights on the one hand, and the defendant's liberty to act on the other.'

[59] *Yearworth v North Bristol* (n 22) [49]: '(c) The unit acquired exclusive possession of the sperm. (d) The unit held itself out to the men as able to deploy special skill in preserving the sperm. (e) Analogously to its admission in relation to the claims in tort, the Trust admits that, if the unit was a bailee of the sperm, it was in breach of the duty of care consequent upon the bailment. (f) The

in *Re: Organ Retention* were required to show that healthcare professionals, by adopting a practice where the details of the post-mortem examination of deceased children are not provided to the parties 'without thought to the individual circumstances of each parent or family', were in breach of their duty of care to the parents.[60]

b. Exclusion, Governance, and the Content of the Primary Duty

Under duties of non-interference, a duty-bearer's permissible actions are determined by an exclusionary boundary. If we accept the set of premises that explain why items of property and the body represent spheres of undefined activity, then it follows that any intrusion into this sphere ought to amount to breach of a duty. The intrusion itself is a wrong thing. There is a basic inequality between the rights-holder and the duty-bearer that follows from the exclusion strategy insofar as 'the interests of [the entitlement-holder] unilaterally determine the contours of what is supposed to be a bilateral relationship of equals'.[61] Coupled with the other structural features of the exclusionary boundary (pre-existing and original rights), the exclusion strategy would appear to create an 'extreme solicitude'[62] for the property rights-holder.

The governance strategy that delineates the rights and duties associated with an activity or interaction can be understood as starting from a position of 'basic equality' between the parties in a bilateral relationship. In order to shift responsibility for the right-holder's loss onto the duty-bearer, the governance strategy requires a finding of fault. That is, some 'wrong doing', or conduct that falls below an expected standard (breaching a duty of care), in addition to causing interference or infringing the interests of the rights-holder. As discussed above, the activity or interaction between the rights-holder and duty-bearer forms the *basis* for why such duties of care are imposed on the duty-bearer. Hence, unlike the basic inequality as between the rights-holder and the duty-bearer that features under the exclusion strategy, the governance strategy posits a 'bilateral relationship of equals' who obtain rights and undertake duties through their participation in an activity, transaction, or interaction.

unit extended, and broke, a particular promise to the men, namely that the sperm "will be stored ... at minus 196°C", quoted at [6(c)(ii)] above.'

[60] *Re: Organ Retention* (n 23) [562].

[61] E Weinrib, *The Idea of Private Law* (Harvard University Press, Cambridge MA 1995) 179–80.

[62] E Weinrib, *The Idea of Private Law* (n 61) 179–80.

c. Complementary Duties and the 'No Property Rule'

As we have already encountered, the exclusion strategy and the governance strategy are overlapping strategies. A duty-bearer therefore may be under two different duties with regards to the same interaction with an object or resource. For instance, an item of property is protected by duties of non-interference under conversion as well as duties of care under negligence. Similarly, trespass to the person prohibits physical contact and negligence imposes liability where a breach of a duty of care causes personal injury. These two types of duties are complementary. On one hand, duties of non-interference address voluntary or 'deliberate' conduct that is inconsistent with the rights of the rights-holder. Negligence, on the other hand, 'covers those situations in which there has been on the part of the defendant no conduct which could be classed as voluntary, and yet which nonetheless interferes with' the rights of the rights-holder.[63]

The 'no property rule' creates a gap in the legal protection of the use and storage of bodily material insofar as there is no basis for duties of non-interference. Such duties that are imposed under the exclusion strategy require inaction or abstinence. Hence, we have seen duties of non-interference employed to protect property rights and rights of bodily integrity. However, because of the 'no property rule' and the judicial unwillingess to extend the rights of bodily integrity to bodily material, there remains only the imposition of duties of care, under the governance strategy, to protect entitlements in bodily material. Rather than being protected by complementary duties of care and non-interference, the current law inadequately protects the use and storage of bodily material by relying upon only the governance strategy.

Before we move on to consider the legal structure of the contingency of property rights, let us consider the attraction and implications of pulling entitlements in bodily material behind the exclusionary boundary. The structure of protection that is afforded to exclusive rights, such as property rights and the right to bodily integrity, includes rights that are exercisable against an open set of persons, rights that vindicate both the original and derivative dimensions of ownership so as to recognize broader categories of actionable loss, and duties imposed on others that include duties of non-interference. As we can begin to appreciate, this exclusionary boundary may be an appropriate approach to the legal protection of entitlements in separated bodily material. At the same time, we can also appreciate that, by employing the exclusionary boundary we allow the rights-holder to 'unilaterally determine the contours'

[63] S Green, 'Understanding the Wrongful Interference Actions' (n 16) 25.

of the legal relationship between the rights-holder and the duty-bearer. This creates a basic inequality between the rights-holder and the duty-bearer. As we will consider in the next chapter, there may be instances where we ought to reduce or mitigate this basic inequality between the rights-holder and the duty-bearer.

IV. The Content of the Remedial Duty

We now turn our attention to the structural features that stem from the conceptualization of property rights as contingent rights. Recall that contingent rights enable and protect preferences and choices that can exist independently of the rights-holder. I will suggest in this section that the law remedies the interference with these rights through the imposition of corrective remedial duties. In comparison, for rights that enable preferences and choices that cannot exist independently of the rights-holder, the law remedies interference with these rights through the imposition of redistributive remedial duties. We are attempting here to identify the *content of the remedial duty*. In other words, we are attempting to identify what losses are recoverable.[64] Even if the claim is actionable per se, we nonetheless still need to identify what losses are recoverable by the claimant against the defendant. Equally, if the claim is actionable upon proof of damage, we nonetheless still need to identify what types of losses qualify as proof of damage.

a. The Principle of Corrective Justice

The principle of corrective justice can be understood as 'the maintenance and restoration of the notional equality with which the parties enter the transaction'[65] or interaction. It follows that corrective justice is concerned with how a wrongful interaction or 'transaction is reversed, undone or counteracted'.[66] According to Ernest Weinrib, this species of justice 'has become central to contemporary theories of private law'.[67]

John Gardner provides a 'continuity thesis' which explains the centrality and dominance of corrective justice in private law.[68] According to the thesis,

[64] D Nolan, 'New Forms of Damage in Negligence' (n 26) 61: '[T]he issue of the damage sufficient to establish a cause of action should not be confused with the harms for which recovery is permitted once the cause of action has been established.'

[65] EJ Weinrib, 'Corrective Justice in a Nutshell' (2002) 52 *UTLJ* 349.

[66] J Gardner, 'What is Tort Law For? Part 1. The Place of Corrective Justice' (2011) 30 *Law & Phil* 1, 9.

[67] EJ Weinrib, 'Corrective Justice in a Nutshell' (n 65) 349.

[68] J Gardner, 'What is Tort Law For? Part 1' (n 66) 27; JL Coleman, 'Corrective Justice and Property Rights' (1994) 11(2) *Soc Phil & Pol'y* 124.

entitlements are allocated by an 'initial norm of justice'. In addition to being a *ground* for allocation (provides a criterion for determining the distribution of the entitlements),[69] the initial norm of justice is a *reason* for the allocation; it provides a normative underpinning for why the entitlements are distributed in a particular way[70] and gives rise to a primary obligation, or a set of primary reasons, to respect the allocation.

Remedies come into play when the norm of justice is not adhered to. According to Gardner, even when the norm of justice is not complied with, the primary reasons that underlie the norm of justice persist so as to still require adherence to the norm:

> While an obligation is either performed or not performed, those reasons in favour of the action that contribute to its obligatoriness can each be conformed to more or less perfectly.[71]

Enter corrective justice. In cases where it is no longer possible to perform the primary obligation 'one can often nevertheless still contribute to the satisfaction of some or all of the reasons that added up to make the action obligatory'.[72] Hence, the remedial obligation arises to rectify the loss caused by the failure to adhere to the primary obligation. For Gardner, 'the normal reason why one has an obligation to pay for losses that are wrongfully occasioned'—to rectify the loss—is that 'it constitutes the best available conformity with, or satisfaction of, the reasons why one had the [primary] obligation'.[73] In a similar vein, Jules Coleman describes this obligation as conforming to the 'local understanding of a more general norm'.[74] Corrective justice, by requiring the reversal of wrongful transactions, reallocates the entitlements back to the arrangement required by the primary obligation (which reflects the allocation determined by the initial norm of justice). If the initial norm of justice requires that Joanna is in position A, and if Lucy then wrongfully causes Joanna to be in position B then the primary obligation on Lucy to respect Joanna-in-position-A still applies and requires Lucy to put Joanna back into position A.

Note that the assumption here is that there are no normative consequences of the interaction beyond the misallocation of entitlements. The initial norm of justice is applicable where it is possible to restore the misallocation of

[69] J Gardner, 'What is Tort Law For? Part 1' (n 66) 27.
[70] J Gardner, 'What is Tort Law For? Part 1' (n 66) 27.
[71] J Gardner, 'What is Tort Law For? Part 1' (n 66) 30.
[72] J Gardner, 'What is Tort Law For? Part 1' (n 66) 33.
[73] J Gardner, 'What is Tort Law For? Part 1' (n 66) 33–4.
[74] JL Coleman, 'Corrective Justice and Property Rights' (n 68) 133.

entitlements, or restore the preferences and choices that the entitlements enabled. In such cases, the remedial obligation can easily be performed and the primary reasons can still be adhered to: since Joanna had a perfectly good bicycle, which Lucy converted or negligently destroyed, the damages Lucy ought to pay is the cost of replacing Joanna's perfectly good bicycle or repairing Joanna's bicycle to its original state. Damages can repair the loss because they are able to substitute for loss by either compensating for the loss in value or restore damaged goods, thereby returning the parties to the initial allocation of entitlements, and adhering to the initial norm of justice.

One of the structural features of property law is the way in which it remedies interference with entitlements by restoring, as near as possible, the initial allocation of the entitlements in the object or resource. This is possible because property rights represent preferences and choices that are contingent to, or can exist independently from, the person. It follows that when an entitlement is interfered with, the law is merely charged with restoring the allocation of entitlements. In essence, the law takes aim at restoring the entitlements, and presumes that the restoration of the preferences and choices that the entitlements enable will follow.

b. Corrective Remedial Duties

Since the remedial aim of property law is corrective, 'special damages' are awarded so that 'the plaintiff is placed, as far as money can do it, in the same place as if the loss had not been inflicted on him'.[75] Accordingly, where property is destroyed or damaged, the reasonable cost of replacement or repair will be recoverable, including any loss of profits from profit earning property. The remedial measure under conversion is subtly different. The full market value of the good is recoverable which is said to reflect 'the action enforcing an involuntary purchase by the converter'.[76]

For example, in *US v Arora,* where a researcher destroyed a cell-line, two types of damages were awarded. The cost of replacement and repair to flasks and laboratory equipment was awarded ($176.78).[77] As a successful action in conversion, the market value of the cell-line was also recoverable. Since it was held that the cell-line had limited extrinsic or market value, the costs of recreating the cell-line were instead awarded ($273.52).[78] The assumption here is that losses suffered from interference can be corrected by damages. This

[75] NE Palmer and E McKendrick (eds), *Interests in Goods* (2nd edn, Lloyd's of London Press, London 1998) 548.

[76] NE Palmer, *Interests in Goods* (n 75) 548.

[77] *US v Arora* (1994) 860 *F Supp* 1091, 1110.

[78] *US v Arora* (n 77) 1110.

is a valid assumption where property law governs contingent rights. Since contingent rights represent preferences and choices that have no particular association with the entitlement-holder, the loss suffered from the interference with the entitlements can be corrected or counteracted with an award of damages.

c. The Principle of Distributive Justice

However, where there are normative consequences beyond the misallocation of entitlements the aims of corrective justice are, in a strict sense, inapplicable. This is because corrective remedial duties, which concern the restoration of entitlements through damages, are unable to address the normative consequences beyond the misallocation of entitlements. In such circumstances, the principle of distributive justice surfaces as the alternative to corrective justice as the remedial aim of the law. The aim of distributive justice is to divide the benefits and burdens among several potential entitlement holders in 'accordance with some criterion that compares the relative merits' of the potential entitlement-holders.[79] When remedies are determined by a distributive justice norm, the remedial response is essentially a second, and potentially *carte blanche*, allocation.

The crucial difference between distributive justice and corrective justice is that corrective justice reallocates the entitlements in an attempt to adhere to the *initial* norm of justice and reverse the wrong done, whereas distributive justice allocates entitlements and responsibilities in accordance with a *new* norm of justice. The reason why a distributive, rather than a corrective, approach is applicable to non-contingent rights is because the restoration of the initial allocation of entitlements is either not possible, or because the restoration of entitlements is unable to redress the preferences and choices that are necessarily associated with the rights-holder. For example, Lucy may be deprived of possession or control of a miscarried foetus or deceased infant, Joanna may be unable to undergo assisted reproductive therapy due to inadequately stored ovum, or the bone marrow donated from Matthew to Mark may be misdirected or destroyed. Corrective justice is inapplicable for two reasons. First, in the case of Lucy and Joanna at least, it is not possible to restore the allocation of entitlements (as required by the initial norm of justice). Second, monetary damages are unable to substitute for preferences and choices that have been forgone because of the interference with their

[79] EJ Weinrib, 'Corrective Justice in a Nutshell' (n 65) 349.

entitlements. Rather, the appropriate remedial response is to award damages in accordance with a new norm of justice.[80]

The new norm of justice (under the principle of distributive justice) may nonetheless resemble, in some regards, the initial norm of justice. As Gardner explains, although the continuity thesis does not apply to irreparable losses, some of the 'implications of the continuity thesis are preserved in the norms regulating general damages'.[81] In other words, although damages for pain, suffering, or distress do not represent an attempt to restore the wronged party to the original pattern of entitlements, there may be a 'corrective spill over'. There is spill over insofar as the loss suffered by the wronged party is relevant in terms of determining the relative merit of the parties in the second allocation, just as the responsibility of the wrongdoer in causing the loss or harm is relevant in determining the merits of the parties in the second allocation. Hence, distributive justice may take on *compensatory* aims.

The new norm of justice is able to embrace a wider range of considerations. As Margaret Radin explains, the aim of the damages may be to express public recognition of the infringing party's fault rather than providing a benefit equivalent to the wronged party's loss:

> Compensation can symbolise public respect for rights and public recognition of the transgressor's fault by requiring something important to be given up by one side and received on the other, even if there is no equivalent of value possible.[82]

Hence the blameworthiness of the infringing party's role in causing the loss and the irreparable loss suffered by the wronged party may be important considerations for determining damages under a distributive justice approach. The remedial duty seeks to allocate a set of benefits and burdens, in response to the wrongful interference, in light of the relative merits of the parties. This is because there are normative consequences above and beyond the mere misallocation of entitlements from the interference with non-contingent rights that the law ought to be able to address.

d. Distributive Corrective Duties

The principle of distributive justice requires a remedial response that is responsive to the normative consequences beyond the mere allocative

[80] J Gardner, 'What is Tort Law For? Part 1' (n 66), 56: 'For the most part, the continuty thesis does not explain general damages. By and large they are not reparative in the strictest sense. They are paid in respect of certain irreparable results of consequences of a tort or breach of contract. As such, they cannot be explained directly by the contintuity thesis. They require a different rationale.'

[81] J Gardner, 'What is Tort Law For? Part 1' (n 66) 1.

[82] MJ Radin, 'Compensation and Commensurability' (1993) 43 *Duke L J* 56.

consequences. In practice, this requires the legal presumption that there are losses that go beyond the misallocation of entitlements and seeks to recognize, and perhaps compensate for, the loss. As Normann Witzleb identifies, torts that concern an individual's personality, such as false imprisonment, assault, and malicious prosecution award ('general') damages for emotional distress. 'As far as these interests are concerned', Witzleb notes:

> the law presumes that the defendant's wrongful conduct will cause some mental distress and claimants may be awarded substantial damages without being required to prove actual emotional harm.[83]

The approach taken under 'general' damages is to presume that there has been a personal loss or suffering following the interference with the entitlements of the rights-holder. General damages are therefore concerned with compensating the normative consequences of interference that go beyond the misallocation of entitlements. For example, consider the 'wrongful conception' in *Rees v Darlington Memorial Hospital NHS Trust*.[84] The majority held that general damages in the form of a 'conventional sum' may be available to the parents to 'compensate' the parents for their loss of 'opportunity to live their lives in the way they wished' and 'limit the size of their family'.[85] According to Nolan, the award 'represents a significant departure from established principles' since the loss of reproductive autonomy 'would no longer be merely a compensatable aspect of a more orthodox form of damage'.[86] Rather, non-pecuniary losses from an unwanted pregnancy, including the loss of reproductive autonomy, are now directly and independently recoverable. The availability of these damages is best understood as being a part of an allocation of benefits and burdens in accordance with a new norm, and cannot be oriented in terms of the application of corrective justice.

Consider, as a further example, Nwabueze's 'remedial quagmire' that arises where an 'organ donated by A for transplantation to B might be intentionally or negligently redirected to C, or maliciously damaged by D'. We are told that organs for transplantation concern an 'element of [physical] deprivation' that 'distinguishes organ transplantation cases from those of pure emotional harm'.[87]

[83] N Witzleb, 'Monetary Remedies for Breach of Confidence in Privacy Cases' (2007) 27 *LS* 430, 446.

[84] *Rees v Darlington Memorial Hospital NHS Trust* [2003] UKHL 52.

[85] *Rees v Darlington Memorial Hospital NHS Trust* (n 84) (Lord Millet) [123], note also the alternative understanding of the award based on Lord Bingham's judgment; that the award 'afforded "some measure of recognition" of the wrong done' (*Rees v Darlington Memorial Hospital NHS Trust* (n 84) (Lord Bingham) [8]). See also D Nolan 'New Forms of Damage in Negligence' (n 26) 79.

[86] D Nolan 'New Forms of Damage in Negligence' (n 26) 79.

[87] RN Nwabueze, 'Remedial Quagmire' (n 46) 206.

Curiously, Nwabueze then uses *US v Arora* as an example of how a proprietary approach should guide the approach to misdirected organs.[88] Yet, *US v Arora* is illustrative of the corrective approach to damages that is inapplicable to remedying interference with non-contingent interests, such as A's and B's interest in the donated organ. Under *US v Arora*, the market value of the cell-line or the cost of re-creating the cell-line was recoverable.[89] Recall the above plight of Lucy (miscarried foetus), Joanna (inadequately stored ovum), Matthew, and Mark (inter vivos bone marrow transplant). Unlike in *US v Arora*, it is not possible to put Matthew, Mark, Lucy, Joanna, 'A', or 'B' in the 'same place as if the loss had not been inflicted' on them. Neither the foetus, the ovum, bone marrow, nor the donated organ, has an extrinsic market value. Nor is it possible to re-create the opportunity that the items of bodily material enabled. As we shall see, even if the costs of re-extraction are recoverable in the bone marrow scenario, Matthew or Mark would still confront additional remedial hurdles under a proprietary approach to compensation for the non-pecuniary loss arising from the misuse of the bone marrow and the need for re-extraction.

The problem with the corrective justice approach is that the law takes aim at restoring entitlements, and presumes that the restoration of the preferences and choices that the entitlements enable will follow. This approach is applicable where the entitlements in bodily material enable preferences and choices that are contingently associated with the entitlement-holder so that the loss suffered is commensurable with monetary damages. Yet, where the entitlements in bodily material enable preferences and choices that are particular to the entitlement-holder, the award of 'general damages' is the appropriate remedial response.

e. Additional Hurdles under Corrective Remedial Duties

That is not to say that non-pecuniary loss is not recoverable under the corrective remedial duties employed by property law. But note the further remedial hurdles that a claimant is expected to clear. Non-pecuniary loss may be recoverable under a property rights-based action where a claimant is able to establish that they suffered a psychiatric injury or emotional distress.[90]

[88] RN Nwabueze, 'Remedial Quagmire' (n 46) 208.

[89] *US v Arora* (n 77) 1110.

[90] As JK Mason and GT Laurie note in 'Consent or Property? Dealing with Body Parts in the Shadow of Bristol and Alder Hey' (2001) 64 *MLR* 710, 728: 'The action for negligence can also provide a remedy to relatives who have suffered nervous shock—but this requires evidence of psychiatric disease, not mere upset or hurt to feelings' which they compare with the 'US position where damages for infliction of emotional distress and negligent handling of a corpse are recoverable by relatives if the body of a deceased family member is mistreated: *MacKay v US* (1993) 8 F 3d 826 and *Gonzalez v Metro Dade City Health Trust* (1995) 651 So 2d 673.'

Otherwise, '[e]motional responses to unpleasant experiences of even the most serious type do not found a claim for damages'.[91]

In order to succeed in a claim for psychiatric injury, two particular elements need to be established. 'The first hurdle', according to Lord Bridge in *McLoughlin v O'Brian*, 'is to establish that [the loss suffered is], not merely grief, distress, or any other normal emotional, but a positive psychiatric injury'.[92] This means that the loss suffered must satisfy either the American Diagnostic and Statistical Manual of Mental Disorders or the International Statistical Classification of Mental and Behavioural Disorders.[93]

The second hurdle is foreseeability. Under negligence, an award for damages in respect of a psychiatric injury is only recoverable subject to proof of causation and foreseeability.[94] Similarly, the damages awardable for breach of bailment are governed by contractual principles that require the non-commercial loss be a 'a major or important part of the object'[95] of forming the bailment relationship. This requirement, as applied by the Court of Appeal in *Yearworth v North Bristol NHS*, amounts to a requirement that the psychiatric injury be a 'foreseeable consequence upon the breach of duty'. This a standard that the claimants in *Yearworth v North Bristol NHS* were able to satisfy.[96] In contrast, in *Re: Organ Retention*, although Mrs Harris succeeded in establishing that the healthcare professionals owed her a duty of care, and that duty was breached, the claim in negligence failed since it was held that the psychiatric injury that she suffered was not foreseeable.[97]

If a loss cannot satisfy the criteria for identifying a psychiatric injury, damages for emotional distress may be recoverable. Yet, 'English law does not

[91] *RK v Oldham NHS Trust* [2003] Lloyd's Rep Med 1 (QB) [20]; R Mulheron, 'Rewriting the Requirement for a "Recognised Psychiatric Injury" in Negligence Claims' (2012) 32 *Oxford J Legal Stud* 77. See also D Nolan, 'New Forms of Damage in Negligence' (n 26) 61: '[A] distinction should be made between forms of harm which are never actionable in negligence, and those which are actionable only in certain limited circumstances. Emotional harm falling short of a recognised psychiatric illness is in the former category; recognised psychiatric illnesses themselves are in the latter (along with pure economic loss).'

[92] *McLoughlin v O'Brian* [1983] 1 AC 410 (HL) 431; R Mulheron, 'Rewriting the Requirement for a "Recognised Psychiatric Injury" in Negligence Claims' (n 91) 80.

[93] R Mulheron, 'Rewriting the Requirement for a "Recognised Psychiatric Injury" in Negligence Claims' (n 91) 84.

[94] *Attia v British Gas* [1988] 1 QB 304; *Yearworth v North Bristol NHS* (n 22) [55].

[95] *Farley v Skinner (No 2)* [2001] UKHL 49, [2002] 2 AC 732 (Lord Steyn) [24].

[96] *Yearworth v North Bristol NHS* (n 22) [54]: '[W]hether for the purposes of his claim in tort or in bailment, each man indeed needs to establish—as set out at [52(d)] above—that his psychiatric injury or distress was a reasonably foreseeable consequence of the breach of duty.'

[97] *Re: Organ Retention* (n 23) [259]: 'Mrs Harris is a primary victim. She is not precluded from bringing a claim against the defendants for psychiatric injuries in negligence. But because I have found that the psychiatric injury suffered by her was not foreseeable her claim does not succeed.'

currently recognise the infliction of emotional distress as a separate cause of action'.[98] Rather a claimant must satisfy the elements of another action and then attempt to add to the damages an award for emotional distress.

There are two notable features about the way in which property law addresses non-pecuniary loss. First, the claimant must prove a causal link between the loss and the breach of duty, and prove that the loss suffered was foreseeable. Second, unless the claimant can satisfy the requirements of a psychiatric injury, damages for emotional distress alone are not recoverable as a standalone category of loss. Where a personal interest is harmed, in order to obtain a remedy under property law, claimants must satisfy the additional elements of—what is treated as—an extra-ordinary remedial measure.

f. Remedial Duties and Bodily Material

The distinction here is between the corrective remedial duties under property law that award 'special damages' and the distributive duties and the award of 'general damages' under torts that are concerned more with personal rights and interests. The corrective aim of property law presumes that there are no normative consequences to interference with entitlements beyond the misallocation of entitlements. The initial norm of justice can be adhered to by restoring the original allocation of entitlements. This was the approach taken in *US v Arora*. Non-pecuniary damages are recoverable where the loss is a foreseeable consequence of the breach of duty. Such losses were proved in *Yearworth v North Bristol NHS*, but limited the success of the claims in *Re: Organ Retention*. For non-contingent rights, it may not be possible to adhere to the initial norm of justice. In such instances, the remedial aim ought to be able to compensate the loss suffered by the wronged party beyond the misallocation of entitlements. Legal actions that protect non-contingent rights, such as battery, defamation, or breach of confidence, impose redistributive remedial duties and the non-pecuniary losses flowing from the infringement of personal rights are directly recoverable.

V. The Transferability of the Right

The 'right to transfer property' has been described as 'an inherent feature of property rights'.[99] Recall that for Stephen Munzer, transferability represents

[98] R Hardcastle, *Law and the Human Body: Property Rights, Ownership and Control* (Hart Publishing, Oxford 2009) 190.

[99] JE Penner, *The Idea of Property in Law* (n 6) 79.

the dividing line between personal rights (such as free speech, privacy, reputation, and bodily integrity) and property rights.[100] The right to transfer entitlements in objects and resources to another person is the fifth and final structural feature of property rights that we will consider in this chapter.

Property rights, as contingent rights, represent preferences and choices that are not associated with any particular rights-holder; Lucy can exercise the same set of preferences and choices in a painting or wine collection than that exercised by Joanna. It follows that the transfer of contingent rights is presumptively permissible since there are no normative consequences from the change in rights-holder. In contrast, 'personal rights', as non-contingent rights, are inalienable. Non-contingent rights represent a set of preferences and choices that can only be fully exercised by the rights-holder; Lucy cannot exercise the same set of preferences and choices in Joanna's bodily integrity or personal information that can be exercised by Joanna.

Despite this distinction between alienability and inalienability there exists an important middle ground. Guido Calabresi and Douglas Melamed[101] have formulated a distinction between three different ways of protecting an entitlement: property rules, liability rules, and inalienability rules. Their framework is based on an economic perspective, which differentiates these different sets of rules in terms of the cost of acquiring an entitlement. The *economic consequences* of acquiring an entitlement identified by Calabresi and Melamed reflect the *normative consequences* of the change of rights-holder as identified by James Penner (see pages 125–8).

As I will explain here, where there are only allocative consequences in terms of the change in entitlements-holder, the parties themselves ought to be able to determine the content of the transaction. In such instances, entitlements are governed by 'property rules'.[102] Where the entitlements enable preferences and choices that can only be exercised by the entitlement-holder, the entitlements cannot be transferred to another person. Such entitlements are protected by 'inalienability rules'.[103] A middle ground emerges where the entitlements are transferable but there may be wider normative consequences to the transaction or interaction (beyond the parties to the transaction or interaction) that require 'an organ of the state' to govern the transaction. In

[100] SR Munzer, *A Theory of Property* (Cambridge University Press, Cambridge 1990) 47–8.

[101] G Calabresi and AD Melamed, 'Property Rules, Liability Rules and Inalienability: One View of the Cathedral' (1972) 85 *Harv L Rev* 1089.

[102] G Calabresi and AD Melamed, 'Property Rules, Liability Rules and Inalienability' (n 101) 1092–111.

[103] G Calabresi and AD Melamed, 'Property Rules, Liability Rules and Inalienability' (n 101) 1111–15.

this third instance, the entitlements are governed by 'liability' or 'governance' rules.[104]

a. Property Rules

According to Calabresi and Melamed, if an individual wishes to acquire an entitlement in a thing from the entitlement-holder, and if the entitlement is protected by a *property rule*, then the individual 'must buy it from him in a voluntary transaction in which the value of the entitlement is agreed upon by the seller'.[105] Transfer under property rules involves a value determined by private actors. Property law permits the transfer of the right since, as a contingent right, there are no normative consequences (beyond mere allocative consequences) that follow from the change in rights-holder in the transaction. The allocative consequences are addressed through the agreement to the exchange. Recall how in chapter three (pages 91–4) the sale of a thing seeks to 'establish a "moral equilibrium" between parties to the transaction'.[106] There is no basis, therefore, to limit or prohibit the free transfer of entitlements in things between persons where the entitlements represent contingent rights since the agreement establishes a 'normative equilibrium between parties to the transaction'.

Once entitlements in bodily material are viewed as property rights, there are few (if any) limits on the transferability of the property right. It is this feature of property law that explains the motivation towards treating stored semen as items of property in the circumstances that arose in *Hecht v Superior Court of Los Angeles County* and *Bazely v Wesley Monash IVF*.[107] As items of property, the right to manage and use the material are transferable rights provided that transfer satisfies the few procedural requirements under the law of intestacy and inheritance. More significantly, property law has been used to enable the transfer of possession of bodily material in order to *narrow* governance or liability rules. In *Re: Edwards*, the court was able to narrow the statutory requirement of progenitor consent (as required under s 21 of the Assisted Reproductive Therapy Act).[108] By finding that the widow of the

[104] G Calabresi and AD Melamed, 'Property Rules, Liability Rules and Inalienability' (n 101) 1092–111.

[105] G Calabresi and AD Melamed, 'Property Rules, Liability Rules and Inalienability' (n 101) 1092.

[106] D Bollier, *Silent Theft: The Private Plunder of Our Common Wealth* (Routledge, New York 2002) 40.

[107] *Bazley v Wesley Monash IVF* [2010] QSC 118; *Hecht v Superior Court of Los Angeles County* (1993) 20 *Cal Rptr* 2d 275.

[108] *Jocelyn Edwards; Re the Estate of the Late Mark Edwards* [2011] NSWSC 478; Assisted Reproductive Technology Act 1997 (Qld Aus), s 21: 'An ART provider must not supply a gamete

deceased had a property right in the semen, the court permitted the widow to assert her property right to the bodily material, rather than interpreting transfer of the semen as an instance of a healthcare institution 'supplying' bodily material (which was prohibited in the absence of progenitor's consent). As the reasoning in *Re: Edwards* demonstrates, claims to use bodily material without the consent of the progenitor may succeed once the material is fitted into a property framework since the framework *enables* the transfer of entitlements in things.

For the same reasons that property rights are transferable, property rights are also divisible. It is because property rights enable preferences and choices that can exist independently of the rights-holder, there are no normative consequences (beyond mere allocative consequences) that follow from property rights being divided.

For example, consider the straws of semen from the same progenitor purchased for $250 per straw by a former couple in *JCM v ANA*.[109] The couple each gave birth to a child using the semen. Upon separation, the former couple disagreed as to the future use of the remaining 13 straws. The Supreme Court of British Columbia held, loosely following the reasoning in *Yearworth v North Bristol NHS*, that the straws of semen were items of property that 'should be divided between the parties' in way that is consistent with 'an agreement to divide their joint assets'.[110] That is, 'to divide their assets fairly' and evenly.[111] Assuming that the 13th straw was indivisible, the court awarded seven sperm straws to the claimant and six sperm straws to the respondent whilst requiring the claimant to pay '$125 for the extra one-half straw she is receiving'.[112] The underlying assumption here is that the only normative consequence of the (uneven) division of the straws of semen is the (uneven) allocation of entitlements. It follows then that the uneven allocation could be off-set by the payment of $125.

b. Governance Rules

In contrast to the permissible framework of property, 'governance rules' (or 'liability rules') represent a restrictive framework for the transfer of entitlements. Calabresi and Melamed's description of liability rules and property rules suggest that they are both permissible frameworks. What their analysis conflates is the difference between legal permission to obtain an entitlement

or an embryo to another person (including another ART provider) except with the consent of the gamete provider and in a manner that is consistent with the gamete provider's consent.'

[109] *JCM v ANA* (2012) BCSC 584.
[110] *JCM v ANA* (n 109) [95].
[111] *JCM v ANA* (n 109) [95].
[112] *JCM v ANA* (n 109) [96].

and the economic costs of obtaining an entitlement. To be 'entitled' to pollute a river, for instance, is governed by a liability rule. These 'liability rules', according to the Calabresi and Melamed, involve:

> an additional stage of state intervention: not only are entitlements protected, but their transfer or destruction is allowed on the basis of the value determined by some organ of the state rather than the parties themselves.[113]

Calabresi and Melamed's economic perspective obscures the fact that a polluter may not legally, or normatively, obtain the entitlement to pollute. Nonetheless, the same structure of state intervention for the transfer of entitlements is prevalent in the way that entitlements in bodily material are legally obtained by healthcare institutions. The cost of acquiring an entitlement in an object or resource, what entitlements can be acquired, and how the transfer is facilitated is determined by the 'governance rule' itself (rather than by private agreement). For instance, the Human Tissue Act 2004 (HTA 2004) provides for the transfer of human tissue through prescribing particular consent procedures[114] and limiting what purposes the transferred material can be used towards.[115] The transfer cannot be on a commercial basis but the progenitor may be compensated for expenses incurred and loss of earnings.[116] Similarly, s 12(1) of the Human Fertilisation and Embryology Act 1990 (HFEA 1990) prohibits the payment of any money 'given or received in respect of any supply of gametes or embryos',[117] yet the HFEA Guidelines provide that 'donors may be reimbursed all reasonable expenses incurred in the UK in connection with donating gametes or embryos' and that 'donors may be compensated for loss of earnings up to a daily maximum of £61.28 but with an overall limit of £750 for each course or cycle of donation'.[118]

Donors of tissue and gametes may therefore receive compensation for extracting and transferring their bodily material. The statutory rules govern who may have entitlements in the material,[119] determines the content of

[113] G Calabresi and AD Melamed, 'Property Rules, Liability Rules and Inalienability' (n 101) 1092.

[114] Under s 5(1) of the HTA 2004 it is an offence to do any activity listed in s 1(1), (2) or (3) without the 'appropriate consent' from the 'person concerned'. Schedule 3 of the HFEA 1990 outlines the necessity for 'consent' and 'effective consent' for the storage or use of embryos or gametes by a licensed authority.

[115] Under s 8(1) of the HTA 2004 it is an offence to use or store tissue for a purpose other than a 'qualifying purpose' (as outlined in Schedules 1 and 2 of the Act). Under s 41 of Human Fertilisation and Embryology Act 1990, it is an offence to perform activities that are not in accordance with a license (under Schedule 2) or that are prohibited by ss 3, 3A, and 4 of the Act.

[116] HTA 2004, s 32(7).

[117] Human Fertilisation and Embryology Act 1990, s 12(1).

[118] Human Fertilisation and Embryology Authority, *Code of Practice* (8th edn, London 2013) [13].

[119] HTA 2004, Schedule 1; HFEA 1990, Schedule 2.

the exchange,[120] and prescribes preconditions for a valid transfer.[121] It is still possible to transfer these entitlements because entitlements in bodily material for donation nonetheless represent preferences and choices that are not associated with the entitlement-holder. Yet, there may be other, wider policy concerns that necessitate that the process of transfer, and the content of the transfer, is governed by a set of restrictive rules. As we have encountered in *ex parte Blood*, the possession, use, or transfer of bodily material may be caught, almost entirely, by the application of governance rules. In terms of the scenario in *ex parte Blood*, the governance rules required written consent for the posthumous use of the progenitor's semen by the deceased widow.[122]

Calabresi and Melamed attempt to explain the employment of liability rules in terms of economic efficiency and distributional outcomes. The most common reason, we are told, for employing a liability rule is that market evaluation of entitlements is too inefficient.[123] This may be true in some contexts. Yet, the governance of the transfer of bodily material is justified with reference to concerns about the transfer, such as the authenticity of consent or the welfare of the transferor: what Calabresi and Melamed term as 'paternalism'.[124]

Note that by limiting how entitlements in bodily material are transferred, and upon what terms, the exercise of the right to transfer and the right to income is reduced from enabling an open set of activities to a closed set of activities. In other words, the effect of governance rules is to permit a particular interaction and transaction, and prohibit a set of alternative transactions that would otherwise be exercisable under property rights. The application of governance rules therefore narrows the ability to transfer an object or resource from an open-ended set of choices to a particular activity or interaction.

c. Inalienability Rules

If the law prohibits someone from acquiring an entitlement, then the entitlement is protected by an 'inalienability rule'. According to Calabresi and Melamed, the use of inalienability rules can be also be analysed in terms of economic efficiency and distributional outcomes. Included within the analysis of economic efficiency of inalienability rules is the concern that there

120 HFEA 1990, s 12; HTA 2004, s 32.

121 HTA 2004, s 1(1)–(3); HFEA 1990, Schedule 3.

122 *R v Human Fertilisation and Embryology Authority, ex parte Blood* [1997] 2 All ER 687, 691.

123 G Calabresi and AD Melamed, 'Property Rules, Liability Rules and Inalienability' (n 101) 1110.

124 G Calabresi and AD Melamed, 'Property Rules, Liability Rules and Inalienability' (n 101) 1113.

may be 'external costs' that 'do not lend themselves to collective measurement in an acceptably objective and nonarbitary way'.[125] In other words, there may be normative consequences to the transaction 'beyond the fact that the ownership has changed'[126] that cannot be measured in monetary or economic terms.

Inalienability rules govern non-contingent rights. Since such rights are necessarily associated with the interest-holder, it is not possible for another person to acquire the right. For example, Matthew's right to free speech and right to privacy, and Mark's reputation and right to bodily integrity, each protect preferences and choices that only Matthew and Mark can exercise respectively. These interests are non-transferable: not because of wider policy concerns about the transfer, but because one cannot exercise the preferences and choices of another. When Matthew is deprived of his free speech or Mark of his bodily integrity, there are normative consequences beyond the misallocation of entitlements; there is a loss that cannot be measured 'in an acceptably objective and nonarbitary way'.

Recall the divisibility of the bodily material in *JCM v ANA* and consider how such an approach would be unavailable to the disputes in *Davies v Davies* or in *Evans v Amicus Healthcare*.[127] The preferences and choices in *Davies v Davies* and *Evans v Amicus Healthcare* (and perhaps also in *JCM v ANA*) concern a conflict between a desire for procreation and parenthood and a desire to avoid parenthood altogether (or a desire to avoid procreation with a particular person). These preferences and desires are non-transferable and indivisible because the preferences and desires cannot exist outside of the particular relationship between a progenitor, their gametes, and the subsequent embryos. Either one party is able to attempt procreation or the other party is able to avoid parenthood. There are, therefore, normative consequences (beyond mere allocative consequences) from dividing the embryos between the parties to the disputes. Moreover, even if the Supreme Court of Tennessee in *Davies v Davies* or the Court of Appeal of England and Wales in *Evans v Amicus Healthcare* were to divide the embryos between the parties to the dispute, there would be normative consequences (for at least the party seeking non-use) beyond the mere uneven allocation of embryos that could not be measured 'in an acceptably objective and nonarbitary way'. Rather,

[125] G Calabresi and AD Melamed, 'Property Rules, Liability Rules and Inalienability' (n 101) 1111.

[126] JE Penner, *The Idea of Property in Law* (n 6) 111.

[127] *JCM v ANA* (n 109); *Davies v Davies* 842 SW 2d 588 (1992); *Evans v Amicus Healthcare and others* [2004] EWCA Civ 727.

the interests of the parties to the dispute each have conflicting but inalienable interests.

Finally, we need to be careful here with the difference between a *functional entitlement* and a *legally enforceable right* when identifying which rights are inalienable. At the most basic functional level, when Joanna gives Lucy a painting, Joanna is transferring a possession entitlement in the painting to Lucy. When Joanna also confidentially discloses personal information to Lucy, she is also transferring a functional entitlement in the information to Lucy. In a functional sense, Lucy is in possession of both a painting and personal information. In a legal sense, Lucy obtains an actionable property right with regards to the painting (the immediate right to possession) from Joanna but does not obtain Joanna's actionable right with regards to the personal information (the right to privacy). This is because Joanna transferred a right in the painting to Lucy, but *cannot* transfer her right in her own privacy. Accordingly, if Joanna were to authorize a health authority to possess her bodily material, she may retain a non-transferable right in the material analogous to her right to privacy.

The fifth structural feature of property law, and the second feature that follows from the contingency of property rights, concerns the transferability of a right. I have suggested here that contingent rights are transferable. Where the content of the exchange is determined by private agreement, the exchange is facilitated through the application of property rules. Alternatively, governance rules may delineate the content and circumstances of transfer. Non-contingent rights are non-transferable since such rights represent preferences and choices that can only be exercised by the rights-holder. Hence, by deeming the stored semen to be items of property, the court in *Bazely v Wesley Monash IVF* was able to identify a legal avenue through which the rights in the semen can be transferred.[128] As a result, such right were vested in the administrator of the estate. By establishing that the widow had a property right in the semen, the court in *Re: Edwards* was able to narrow the legislative restrictions on the 'supply' of gametes.[129] By finding that the purchased semen in *JCM v ANA* were items of property jointly held by the former couple, the court was able to 'divide their assets fairly' and consistently with their separation agreement by allocating the 13 straws of semen between the two parties.[130]

[128] Queensland Succession Act 1981 (Qld), s 8.
[129] Assisted Reproductive Technology Act 1997 (NSW), s 21.
[130] *JCM v ANA* (n 109) [95]–[96].

VI. Chapter Summary

The structure of the exclusionary boundary employed by property law provides stringent protection of entitlements in objects or resources. Behind the exclusionary boundary, the right to possess, use, or control an object or resource is a right that is enforceable against an open set of persons without needing to show any particular legal relationship with the duty-bearer. Such rights are infringed when the relationship of possession, use, or control is interrupted without the need to show any derivative loss, and a duty-bearer is responsible for infringement when the duty-bearer breaches their duty of non-interference.

It is this stringency of protection that can explain why there is pressure on the common law to recognize property rights in bodily material; to bring such valuable things under the aegis of the exclusion boundary. As we have already seen, the exceptions to the 'no property rule' have been motivated by these structural advantages of the exclusion strategy of property rights. The Lockean exception, as formulated in *Doodeward v Spence*, was introduced in order to ground duties against an open set of persons. The application of the Lockean exception in *R v Kelly* imposed a duty on an open set of persons that was therefore inclusive of an artist who appropriated the body parts from the Royal College of Surgeons. The Hegelian exception in *Yearworth v North Bristol NHS* was motivated by the need for the claimants to identify an actionable loss. Beyond these instances of the application of property law, there remains only the imposition of duties of care under the governance strategy rather than complementary duties of care and non-interference. Hence, we can appreciate the benefit of placing entitlements in objects or resources behind the exclusionary boundary; the exclusionary boundary stringently protects the interest we have in the 'use of things'. Yet, we can appreciate how the exclusion boundary creates a basic inequality between the rights-holder and the duty-bearer. By providing such stringent protection, the exclusionary boundary enables the rights-holder to 'unilaterally determine the contours' of the legal relationship between the rights-holder and the duty-bearer.

Property rights are also rights that can exist independently of the rights-holder. Hence, a duty-bearer's infringement of the property rights of the rights-holder is remedied through the imposition of corrective remedial duties that require a reversal or restoration of the misallocation of entitlements. Moreover, such rights are also transferable, either through the agreement of the transferor and transferee, or through the application of governance rules. Both these structural features of property rights presume that the object or resource enables preferences and choices that can exist independently of the rights-holder. Where this presumption is not applicable, a different structural approach is required: distributive remedial duties, that are able to recognize broader categories of non-pecuniary loss, ought to be imposed and the rights in the object or resource ought to be non-transferable.

We can now begin to sense the limits of property law. There may be instances where the full exclusionary boundary creates an intolerable degree of inequality between the rights-holder and the duty-bearer. There may also be instances where we ought not to presume that the rights in the object or resource enable preferences and choices that can exist independently of the rights-holder. I will argue in the next chapter that these limits of property law are visible when we consider the wholesale application of property law to the use and storage of bodily material. This prompts the formulation of a novel legal structure, subtly distinct from the structure of property law, that ought to govern the use and storage of bodily material.

6

The Limits of Property Law

Natalie Evans was diagnosed as having slow-growing cancerous tumours in her ovaries. Prior to surgery to remove the tumours, Evans and her partner Howard Johnson attended a fertility clinic (Amicus Healthcare) and created six embryos from their extracted gametes. As a result of the surgery to remove the tumours Evans could no longer conceive a child without the use of the embryos. Before an embryo could be implanted, the relationship between Evans and Johnson ended. Johnson wrote to Amicus Healthcare asking to have the embryos destroyed.

Ultimately, the embryos were destroyed. As explained earlier (at pages 22–3), the governing legislation required consent from both Evans and Johnson for each stage of creation, storage, and implantation. Both the Court of Appeal and European Court of Human Rights held that this statutory requirement of continued bilateral consent was a justified limitation on Evan's right 'to respect for her private and family life' on the basis that the legislation also had to account for Johnson's right to the same respect.[1]

AW provided sperm to assist CC to conceive through in vitro fertilization. Using the sperm CC conceived she gave birth to twins. Discontent with his access to the twins, AW refused to consent to the release of the remaining four embryos that were formed using his sperm. According to the Court of Queen's Bench of Alberta, AW and CC 'were not in a relationship of interdependence of some permanence'.[2] When AW provided the sperm, 'he extended a courtesy to a friend by giving her a gift of his sperm', and as a result, 'the gift was an unqualified gift given in order to conceive children'.[3] The court then held that 'the remaining fertilized embryos remain [CC's] property'.[4] As CC's property, 'they are chattels that can be used as she sees fit' and 'AW is not in a position to control or direct their use in any fashion'.[5]

[1] *Evans v Amicus Healthcare and others* [2004] EWCA Civ 727 [111]; *Evans v United Kingdom* (2008) 46 EHRR 34 (Grand Chamber) [90].

[2] *CC v AW*, 2005 ABQB 290 [20].

[3] *CC v AW*, (n 2) [20].

[4] *CC v AW*, (n 2) [21].

[5] *CC v AW*, (n 2) [21].

Johnson retained a legally protected interest in his sperm, and subsequent embryos. Johnson's interest was sufficiently robust to outweigh Evan's 'all-or-nothing' chance to conceive a child that was genetically hers. In contrast, AW relinquished any interest that he may have had in his sperm by providing an unqualified gift to a friend with whom he was no longer in an intimate relationship with. This contrast between these two scenarios reflects an ambiguity in bodily material. Bodily material may be the object of an unqualified gift, where the physical separation coincides with a complete alienation of the object from the subject. Alternatively, bodily material may be an aspect of a person's 'private and family life',[6] where physical separation does not sever the connection between the bodily material and subject. In what follows we confront the practical implications of prescribing legal rules that can govern an ambiguous subject matter.

The purpose of this chapter is to identify the particular structure of rights and duties that ought to apply to the use and storage of bodily material. In order to do so we must assess whether rights that arise in items of bodily material are conceptually consistent or inconsistent with property rights. As we can anticipate, whether a right in an item of bodily material is conceptually consistent with property rights will depend on whether the physically separated bodily material is alienated from, or remains connected with, the progenitor.

We know from the previous discussion that property law employs an exclusion strategy. In the vacuum created by the 'no property rule', the law has relied upon a legislative governance strategy to allocate rights and duties to participants in an interaction that involves bodily material. As I will explain, sole reliance of the governance strategy provides inadequate protection of the entitlements that may arise in bodily material. Moreover, the entitlements that may arise in bodily material often represent interests in an open-ended set of activities, and such interests ought to be protected through the exclusion strategy. Although legislative governance is an important aspect of the overall response of the law to the practice of extracting, storing, and using bodily material, such governance rules are best interpreted as limiting, rather than conflating, the right-holder's sphere of undefined activity that is protected under the exclusion strategy.

Property rights are also rights that enable preferences and choices which can exist independently of the rights-holder. When rights in bodily material arise on the basis that the bodily material retains a functional unity with the body (or continues to function as the medium of social

[6] European Convention on Human Rights (ECHR), Article 8.

experience) such rights are conceptually inconsistent with property law. As I will explain, this conceptual inconsistency reveals the structural limit of property law. In essence, if property law is to protect preferences and choices that cannot exist independently of the rights holder, then the structure of property law will be too restrictive when it comes to remedial duties and too permissible when it comes to the transferability of the right. Alternatively, when the bodily material enables preferences and choices that can exist independently of the rights-holder, rights in such items of bodily material are conceptually and structurally consistent with the application of property law.

Throughout the discussion so far I have relied upon two points of reference: the body and property. My suggestion has been that bodily material is similar to the body insofar as it can perform the same physiological or social function, and bodily material is similar to property insofar as it is an object or resource that can be possessed, used, and controlled. As we narrow our focus in this chapter on the appropriate set of rights and duties that ought to apply to the use and storage of bodily material, I will introduce a further analogy: the rights that we ought to retain in our own bodily material are conceptually and structurally similar to the rights we exercise over our own personal information. The overall contention here is that for bodily material that is *in-itself*, property rights represent the appropriate set of rights and duties. For bodily material that is *for-itself* or *for-others*, the structure of rights and duties that protect the common law right to privacy represent the most appropriate legal response.

I. Exclusion, Governance, and Bodily Material

As we have seen in the previous chapter, there is a lot to be gained from pulling our rights of use and control in an object or resource behind an exclusionary boundary. Yet, as I will explain in this section, the legal response to the use and storage of bodily material has so far relied upon the allocation of legislative rights and duties. This sole reliance on this governance strategy provides an inadequate structure of protection rights of use and control in bodily material. I will suggest that the exercise of rights of use and control over bodily material will align with the three premises that underlie the exclusion strategy. Such rights, therefore, ought to be exclusive rights. This suggestion does not, however, sideline or negate the role of existing legislative rights and duties. Rather, legislative controls on the use and storage of bodily material can be interpreted as limiting rather than deflating the sphere of undefined activity under the exclusion strategy.

a. The Limits of the Governance Strategy

Given any unease that we may have with the application of property law to the use and storage of bodily material, it is tempting to advocate for the statutory allocation of rights and duties as the appropriate legal response to the use and storage of bodily material. For instance, Jonathan Herring argues for a statutory regime on the basis that 'relying on statute gives us a starting point of unbiased scales with which to compare the different individual and social claims that may arise'.[7] Such legislative schemes aim to keep some regulatory control over practices that may raise social or ethical concerns, whilst also enabling patients to have access to emerging treatments and enabling researchers to have access to material resources. These legislative schemes, for the most part, have so far filled the void left by the 'no property rule'.

However, these statutory schemes prescribe rights and duties that arise in the context of particular activities or interactions. The focus of the legislative provisions is invariably on the requirement that consent is obtained for an activity that uses bodily material, and the requirement that the activity represents a procedure that is approved by the relevant regulatory authority. Provided that the activity satisfies these requirements, the activity is a lawful activity. For example, the progenitor's rights of use and control over the fertilized embryo in *Evans v Amicus Healthcare* were facilitated by the requirement of bilateral consent for each act of creation, storage, and use.[8] Without the consent of the progenitors, an act of creation, storage, or use would be unlawful. The retention of the organs of deceased infants in *Re: Organ Retention* was lawful because the parents had consented to a post-mortem.[9] Following England's organ retention controversies, the focal point of the updated legislative scheme remains the right to provide or withhold consent for the retention of bodily material for particular purposes. These statutory schemes govern *particular activities.* This strategy, that apportions rights and responsibilities between participants in an action or activity, is typical of, if not inherent to, statutory schemes.

The governance strategy has limits. Consider, for example, the statutory governance in *Yearworth v North Bristol NHS*. The chemotherapy patients in *Yearworth v North Bristol NHS* had statutory rights to use and manage

[7] J Herring, 'Why We Need a Statute Regime to Regulate Bodily Material' in I Goold et al, *Persons, Parts and Property: How Should We Regulate Human Tissue in the 21st Century?* (Hart Publishing, Oxford 2014) 215.

[8] *Evans v Amicus Healthcare* (n 1) [36].

[9] *Re: Organ Retention Group Litigation* [2005] QB 506 [119]–[127]; Human Tissue Act 1961, s 2(1), (2).

their stored semen. These rights were facilitated by the duties imposed on the fertility unit as a licensed authority. The statutory duties on the fertility unit required that the unit was 'unable to store the men's semen without their consent',[10] 'unable to use it for the purposes of any treatment of persons other than the men themselves (with their wives or partners) without their consent to such use',[11] and unable to store or use the semen if the men withdrew their consent.[12] The focus of the law on particular activities—consent, the withdrawal of consent, use only by the progenitors—narrows the protection of the law to a particular set of activities. The exclusion strategy, in comparison, focuses on the thing itself (the semen, embryo, child's body), and protects an undefined set of activities. Simply put, the main limit of the governance strategy is the narrow scope of protection that it affords only to particular stipulated activities. In *Yearworth v North Bristol NHS* the inadequate storage of the semen was not a breach of a statutory duty and thus the progenitors' rights were unable to be vindicated through the statutory governance of their rights. In order to vindicate the progenitors' rights of use and control in the semen, the Court of Appeal in *Yearworth v North Bristol NHS* introduced property rights, and the accompanying exclusionary boundary, to provide the appropriate structure of protection.

When compared to the exclusion strategy, the governance strategy provides limited legal protection for the *use of things*. Even if a healthcare institution has a right to possess bodily material, the statutory permission or license to possess does not impose a duty on the rest of the world. A third party, who obtains possession of the bodily material, may commit a regulatory offence (by not satisfying the licensing or consent requirements) but does not breach a duty owed to the person with the lawful right of possession. It was the recognition of property rights in *R v Kelly* that was used to broaden the scope of the Royal College of Surgeons' right to possess the preserved body parts beyond their statutory rights.[13] J Kenyon Mason and Graeme Laurie have identified further inadequacies of the governance strategy or 'consent paradigm'.[14] According to their analysis, reliance upon consent as the focal point of the legal relationship constrains the scope of the law in terms of the limited civil remedies available as sanctions for non-adherence with the statutory

[10] *Yearworth v North Bristol NHS Trust* [2009] EWCA Civ 37 [44]; Human Tissue Act 2004 (HTA 2004), Schedule 3, Paragraph 8(1).

[11] *Yearworth v North Bristol* (n 10) [44]; HTA 2004, Schedule 3, Paragraph 5.

[12] *Yearworth v North Bristol* (n 10) [44]; HTA 2004, Schedule 3, Paragraph 4.

[13] *R v Kelly* [1998] 3 All ER 741, 750.

[14] JK Mason and GT Laurie, 'Consent or Property? Dealing with Body Parts in the Shadow of Bristol and Alder Hey' (2001) 64 *MLR* 710, 727.

provisions,[15] the limited circumstances under the common law where consent or authorization is required with regards to interaction with bodily material,[16] and the unclear form of legal protection provided to the right to possess as the corollary to the duty to inter.[17]

b. The Exclusion Strategy and Bodily Material

Sole reliance on the governance strategy struggles to grasp the fullness of the rights of use and control that ought to be able to be exercised in separated bodily material. In comparison to the governance strategy under the legislative schemes, our relationship with items of property and our relationship with our bodies represent the exercise of *exclusive* rights of use and control. Property law, as I have explained earlier (at pages 114–21), is concerned with the use of things (land and toothbrushes), rather than the activities (growing wheat or dental hygiene) that a thing may be involved in. The focus of property law is therefore on our 'purposeful use of things'. Given that the 'purposeful use of things' enables an indefinite list of activities, the law protects the use of things through the exclusion of others. The right to bodily integrity is conceptually and structurally similar to property rights in this regard. The right to bodily integrity is concerned with the use and control of the body itself, rather than the activities (such as mole removal, rugby tackles, or sexual intercourse) that the body may be involved in. The right to use and control our bodies follows from the close (from the third-person point of view) or integrated (from the first-person point of view) relationship between our subjectivity and our bodies. The use of our bodies creates an open-ended, or undefined, set of the activities that the law protects through the exclusion of others.

The right to privacy is also an exclusive right. As Graeme Laurie explains 'informational privacy focuses on the control that an individual can exercise over his or her personal information'.[18] The focal point of the right is the relationship of control over the object or resource (the personal information), rather than the focus being on the activity of the disclosure of personal information to another person nor the various activities that the right to privacy seeks to protect. Although our right to privacy protects a series of activities

[15] JK Mason and GT Laurie, 'Consent or Property' (n 14) 727.

[16] JK Mason and GT Laurie, 'Consent or Property' (n 14) 727.

[17] JK Mason and GT Laurie, 'Consent or Property' (n 14) 727.

[18] GT Laurie, 'Challenging Medical-Legal Norms: The Role of Autonomy, Confidentiality, and Privacy in Protecting Individual and Family Group Rights in Genetic Information' (2001) 22 *J Legal Med* 1, 30.

(such as controlling which aspects of our individuality we share with others or make available for public judgment or criticism)[19] such activities are legally protected through our exclusive control over our personal information. Insofar as 'privacy is defined as a state of separateness from others',[20] the right to privacy carves out a sphere, intrusions into which are prohibited. As we shall see below (at pages 193–6), the legal structure of duties of confidentiality have been adjusted to provide exclusionary protection for a sphere of informational privacy.

Property rights, the right to privacy, and the right to bodily integrity are rights that share the same three conceptual premises: the entitlement-holder has an interest in 'purposely dealing with things',[21] the exercise of the entitlements thereby enables an open-ended set of activities which the law protects through the exclusion of others from the object or resource. With reference to the same three premises that underlie the exclusion strategy, we can now assess whether our legal relationship with bodily material ought to be an exclusionary legal relationship.

In chapter two I argued that a progenitor may retain the right to use and control his or her bodily material where the bodily material retains a functional unity with the body (and thereby continues to be the medium of our being in the world). Such rights of use and control can be seen as an extension of the progenitor's rights of self-ownership. Just as the right to bodily integrity (that is premised upon the principle of 'self-ownership') enables an open set of activities with regards to the body, bodily material that shares a functional equivalency with the attached body (bodily material that is *for-itself*) also enables a range of open-ended choices as to how the material may be used. As we have seen with regards to property rights, privacy rights, and rights of bodily integrity, the way in which the law protects a range of open-ended choices as to how an object or resource is used is through the exclusion of others from the object or resource. The exclusion of others preserves a 'negative liberty' for an undefined set of activities.[22]

Similarly, when partners, parents, spouses, or siblings exercise their rights over the body or bodily material of a deceased person, such rights follow from the recognition that the *relationship with* the body and bodily material

[19] GT Laurie, 'Challenging Medical-Legal Norms', (n 18) 27–8.

[20] GT Laurie, 'Challenging Medical-Legal Norms', (n 18) 27: '[O]ne can deduce a single unifying definition: privacy as a state of separateness from others. Thus, privacy should be taken to refer to a state in which an individual is separate from others, either in a bodily or psychological sense, or by reference to the inaccessibility of certain intimate adjuncts to their individuality and personality, such as personal information.'

[21] JE Penner, *The Idea of Property in Law* (Oxford University Press, 1997) 71.

[22] JE Penner, *The Idea of Property in Law* (n 21) 73.

of the deceased is valuable. As I have explained (at pages 63–4), the body of the deceased remains as a medium of social experience. As with any relationship, the value of the social experience cannot be reduced into a particular activity or interaction. The exercise of entitlements in the body *for-others* is concerned with a relationship with a thing that cannot be reduced to the predictable set of activities. The rights of possession and control of the body therefore represent a range of open-ended choices that ought to also be protected through the exclusion of others from the body or bodily material.

Healthcare institutions, in their therapeutic, scientific, and educational use of bodily material could also engage in purposely dealing *with the thing* (rather than any particular activity or interaction). The material resource could be used towards a range of socially valuable activities, from diagnoses and clinical audits, to education, training, medical treatment, and scientific research. Even in the context of scientific research, the various inquiries that samples of bodily material could contribute to is open-ended. More generally, we recognize that healthcare institutions are able to possess, use, and control bodily material on the basis of the socially valuable outcomes that follow. The more open-ended the set of activities that such institutions can engage in the greater scope for the research, clinical, and therapeutic benefits that follow from the institutional use of bodily material. Recall Henry Smith's cost-based analysis of the exclusivity of property rights. The 'exclusion strategy' operates by having clear boundaries around items of property.[23] Within the boundary 'owners have open-ended choices of how to invest in or consume the asset'.[24] This strategy is said to be more economically efficient than the 'governance strategy' under 'liability rules' because of the greater costs involved in identifying the activities that fall into the scope of permitted use and identifying the loss of value recoverable from interference with particular activities.[25] Healthcare institutions therefore ought to obtain exclusive rights which protect an open-ended set of activities in bodily material (that is *in-itself*).

As Larissa Katz[26] and James Penner[27] have both identified, an exclusive right does not entail the absolute exclusion of all others. There remains logical space for the recognition of subsidiary and reversionary interest holders.

[23] HE Smith, 'Property and Property Rules' (2004) 79 *NYU L Rev* 1719, 1753: 'For reasons of information cost it is often advantageous and almost inevitable that rights will be delineated by means of what I have called an 'exclusion strategy'. Such a strategy relies on rough and low-cost signals that are not tied to use in order to protect indirectly a large and unspecified set of uses.'

[24] HE Smith, 'Property and Property Rules' (n 23) 1728.

[25] L Katz, 'Exclusion and Exclusivity in Property Law' (2008) 58 UTLJ 280.

[26] L Katz, 'Exclusion and Exclusivity in Property Law' (n 25) 289.

[27] JE Penner, *The Idea of Property in Law* (n 21) 74.

The exercise of an exclusive right amounts to exercise of authority over who may and may not interact with the thing. Exclusive rights function 'like a gate, not a wall'.[28] A rights-holder may exercise his or her exclusive rights of use and control to include another person into the sphere of activity. It is conceptually consistent, for instance, to say that the progenitors in *Bazely v Wesley Monash IVF* and *Hecht v Superior Court of Los Angeles County* had an exclusive interest in their stored semen but their partners may nonetheless each have had a reversionary interest in the semen. Co-participation in assisted reproductive therapy may represent a gate through which another (reversionary) entitlement-holder may enter.

Rights in bodily material can therefore be appropriately conceptually characterized as representing exclusive rights, and therefore ought to benefit from the structural protection afforded by the exclusionary boundary. Note that this contention, that our rights in bodily material ought to be viewed as exclusive rights, is contestable. For instance, Jonathan Herring and Pak-Lee Chau suggest, by using a large range of physiological facts, that each body is physiologically interconnected with, and interdependent on, other bodies. Each body, according to Herring and Chau, is constantly interacting and dependent upon the wider environment for survival. Their conclusion is that 'our bodies are not just ours'.[29] Instead, 'our bodies are leaky'.[30]

There is no denying the physiological facts that the Herring and Chau analysis draws upon. However, the view formulated here is that, regardless of how fluid or leaky our physiological existence may be, the physiological facts about us at any one point in time amount to the location, or site, of our subjectivity. Our fluid and leaky bodies are nonetheless the medium through which we exercise our autonomy, experiences preference satisfaction, and form relationships (or however our subjectivity is expressed). From a strictly physiological or third-person point of view, our bodies may not just be ours; they may be contingent and fluid facts about us. However, in terms of the features or attributes that ground duties towards others, *our own bodies* are the focal point of such features or attributes. It is because the body is the site or location of our subjectivity that we have an interest in purposely dealing with our bodies, and an interest in purposely dealing with our bodily

[28] JE Penner, *The Idea of Property in Law* (n 21) 74: 'The right to property is like a gate, not a wall.'

[29] J Herring and P-L Chau, 'My Body, Your Body, Our Bodies' (2007) 15 *Med L Rev* 34.

[30] J Herring and P-L Chau, 'My Body, Your Body, Our Bodies' (n 29) 34, a term coined by Shildrick; M Shildrick, *Leaky Bodies and Boundaries: Feminism, Postmodernism and (Bio)ethics* (Routledge, London 1997).

material that retains a functional unity with the body. This interest, I have argued, ought to be characterized as an exclusive interest.

c. Governance and Exclusion

However, the exclusion strategy, and the structure of the exclusionary boundary, need not be at the expense of the governance strategy. There are numerous instances where the stored bodily material has a narrow range of practical application, such as organs removed for transplantation, tissue retained for diagnosis, or gametes stored for assisted reproductive therapy. Moreover, legislation may limit and regulate the activities that can be lawfully undertaken with regards to the bodily material. For instance, human tissue may only be retained and used towards specified purposes and the use of human gametes and embryos may be lawful under certain regulatory conditions. These functional and legal limits circumscribe, but do not conflate, the undefined sphere of activity.

In *Yearworth v North Bristol NHS*, for example, the Human Fertilisation and Embryology Act 1990 (HFEA) restricted who could store sperm for reproductive treatment and the circumstances in which it could be used for treatment. Yet, as Sarah Green explains 'the Court of Appeal took the view that the limits imposed by the HFEA upon the claimants' right to use their stored sperm were not sufficiently extensive to prevent their having property rights in that sperm'.[31] Beyond the series of legislative rights and duties, the court identified that the claimants also exercised rights arising from their relationship with the thing itself. According to Green:

> Although the Act restricts the ability to use sperm removed and stored for reproductive treatment to those licensed by the Act to do so, thereby preventing the claimants themselves from so using it, the court gave much weight to the negative control of the claimants; that is, the fact that their sperm could neither continue to be stored, nor used for any purpose, without their consent.[32]

The control that an exclusive rights-holder may exercise over his or her bodily material is therefore a 'negative' control over the bodily material. Legislation may therefore limit who may possess bodily material, limit when a person may use bodily material, and prescribe the circumstances in which a person can transfer bodily material. Such provisions narrow or impinge upon the sphere of activity that a rights-holder may pursue. Yet, provided that these statutory provisions do not represent an attempt to completely or

[31] S Green, 'The Subject Matter of Conversion' [2010] *JBL* 218, 239.
[32] S Green, 'The Subject Matter of Conversion' (n 31) 239.

exhaustively govern all activities involving bodily material, there remains conceptual space for the rights-holder to engage in an undefined, albeit limited, sphere of activity.

There are sound reasons for why legislation and regulations limit what can be undertaken with regards to bodily material and provide a series of conditions under which any activities are legally permissible. The challenge for the law is to strike the appropriate balance between two strategies. On one hand, the law ought to govern and restrict the various activities that involve the removal, retention, and use of bodily material. On the other hand, rights in bodily material stem from the rights-holder's interest in the purposeful *use of things*, and therefore protect—to some degree—the open-ended use of bodily material.

Property rights, the right to bodily integrity, the right to privacy, and the rights that arise in bodily material all share the conceptual premises that underlie the exclusion strategy. As a result, property rights, the right to bodily integrity, and (as we shall see below at pages 191–205) the right to privacy are protected by the threefold structure of the exclusionary boundary. Such exclusive rights are nonetheless subject to the limits and restrictions of statutory schemes. The implication, which we will explore below, is that the rights that arise in bodily material ought to also adopt the structure of the exclusion boundary on the basis of the conceptual alignment of such rights with the exclusion strategy.

The *body-as-property* metaphor is therefore a mixed metaphor. We can appreciate the attractive elements of the metaphor. Like our items of property, our bodies *are ours*, and we are the sole arbiters as to what can be done with them. The law then protects this relationship through an exclusionary boundary. I have suggested that our rights of use and control over bodily material ought to also be conceptualized as exclusive rights. When we find ourselves attracted to the property paradigm as the solution to the uncertain legal status of bodily material, it is most likely because we are attracted to the exclusion strategy that property law employs. At the same time, we can also appreciate that our relationship with our body is unlike the relationships that we have with our items of property. We shall turn to consider how the contingency of property rights aligns with the rights that arise in bodily material.

II. Contingent Rights and Bodily Material

In chapter four we assessed whether our right to bodily integrity, that is based upon the principle of self-ownership, is a contingent or non-contingent right. I argued that the right is a non-contingent right since our relationship with our own bodies enables preferences and choices that can only be exercised by us as the self-owner. We now turn to consider whether rights of use and control over bodily material represent contingent or non-contingent rights. In this section I will consider how bodily material is ambiguous insofar as it is capable of being *for-itself*, *for-others* or *in-itself*. I will suggest that the rights of use and control over bodily material that is *in-itself* are contingent rights and thereby share a conceptual consistency with property rights. However, rights of use control over bodily material that is *for-itself* (or *for-others*) are non-contingent rights. There is, therefore, a conceptual inconsistency between rights in bodily material (that is *for-itself* or *for-others*) and property rights. As we will consider in the following section, this conceptual inconsistency creates structural problems for the application of property law to the use and storage of bodily material.

a. Bodily Integrity, Property, and Privacy

Our bodies and our items of property differ conceptually. I explained this difference in terms of preferences and choices that either can or cannot exist independently of the rights-holder. According to the first person point of view developed in chapter two, our subjectivity (however constituted) can be understood in terms of 'tasks'. These tasks represent the way in which our subjectivity is interwoven or integrated with the body. Since we are unable to separate a person's subjectivity from the body, there is a necessary (non-contingent) relationship between a person and their body. Our rights of exclusive use and control over our bodies are therefore non-contingent rights. From the alternative, third-person point of view, the structures and capabilities of a person's body are only necessarily associated with the person only to the extent that the body is closely and immediately connected with the subjectivity of the person. Both views are able to explain why our exclusive rights of use and control over our *bodies* are non-contingent rights.

Our right to privacy is a non-contingent right. To the extent that an 'individual' or subject is 'separate from others', there are 'certain intimate adjuncts to their individuality and personality' that ought to be 'inaccessible' to others.[33] There are various reasons for why we ought to be able to control information that pertains to us as individuals. The ability to control our personal information enables us to exercise our preferences and choices as to which aspects of our individuality we share or relinquish with others,[34]

[33] GT Laurie, 'Challenging Medical-Legal Norms' (n 18) 27.

[34] C Fried, *An Anatomy of Values: Problems of Personal and Social* (Harvard University Press, Cambridge 1970) 142; 'Intimacy is the sharing of information about one's actions, beliefs, or emotions which one does not share with all and which one has the right not to share with anyone.'

control which aspects of our personality that we make available for public judgment and criticism,[35] and determine how we construct a public personae that may be distinct from the realities of our own personhood.[36] These preferences and choices, that underlie the notion of privacy, are necessarily associated with the person to whom the information pertains to. With the right to privacy, as with rights of self-ownership, another person cannot stand in the same position as the rights-holder with regards to the personal information. There are normative consequences from the infringement or invasion of the rights-holder's informational privacy beyond the mere allocation of information to another person. The privacy right-holder loses the value of having control over his or her public personae, and this value cannot be obtained by whoever infringed the right to privacy.

Property rights, in comparison, fundamentally differ from the right to bodily integrity and the right to privacy. No item of property is necessary or essential to our existence, experience, or engagement in the world. Nor does our control over our property necessarily pertain to our individuality or personality. We may find items of property useful, they may to some extent express our likes, dislikes, experiences, and aspirations, but insofar as they are useful to our experience or engagement with the world, each item of property is ultimately substitutional. For every item of property, 'a different person who takes on the relationship to the thing' is able to 'stand in essentially the same position as the first person'.[37]

b. Preferences, Choices, and Bodily Material

The separation of bodily material from the person ends the close and immediate connection between the bodily material and a person's subjectivity. A person may nonetheless retain, or another person may gain, rights of control and use over the bodily material. The exercise of these rights will enable a set of preferences and choices. The key question becomes whether these preferences and choices are necessarily or contingently associated with the rights-holder.

It follows from the third-person point of view that the only particular connection between the person and their body is the close physical connection. Once the physical connection is severed, the bodily material—and the preferences and choices that the bodily material enables—are only

[35] S Benn, 'Privacy, Freedom and Respect for Persons' in F Schoeman (ed) *Philosophical Dimensions of Privacy: An Anthology* (Cambridge University Press, Cambridge 1984) 230–1; K Greenawalt, 'Privacy and Its Legal Protections' (1974) 2 *Hastings Centre Stud* 45.

[36] GT Laurie, 'Challenging Medical-Legal Norms' (n 18) 28.

[37] JE Penner, *The Idea of Property in Law* (n 21) 113.

contingently associated with the progenitor. Another person can, according to this view, stand in the same position with regards to the bodily material as the progenitor. Furthermore, according to this view, there are no normative consequences (beyond mere allocative consequences) to the change in rights-holder. If we adopt this view point, the rights of use and control that a person may retain or gain in bodily material are contingent rights and are conceptually consistent with the application of property law.

The body, according to the first-person point of view advocated here, is the site or location of our subjectivity. The physical separation of bodily material from the body may or may not change the status of the bodily material. Here we encounter again the ambiguities in the body. Where the bodily material continues to be unified with the body in terms of a person's tasks, the bodily material remains functionally equivalent to the body. The bodily material, like the body, is *for-itself*. Alternatively, where the bodily material is no longer 'directed towards an existing or possible task' it becomes like any other object in the world, and has only a 'mechanical' or 'instrumental' relationship with the progenitor. Such items of bodily material, like other objects, are objects *in-itself*. The ambiguity is that bodily material may therefore be *for-itself* or *in-itself*.

I have suggested that the way in which this ambiguity can be resolved is to view a person's body, and his or her bodily material, as self-ascribed. Since a person determines his or her own 'tasks' (determines his or her rationally motivated standards, desires, preferences, plans, intentions, goals, etc.) a person also determines whether an aspect of his or her body or bodily material is directed towards any of the tasks. Where the bodily material is ascribed as retaining a functional unity with the body, and thereby continues to be *for-itself*, the use and control of bodily material (like the use and control of the body) enables a set of preferences and choices that can only be exercised by the progenitor. The rights of use and control over our body and our bodily material that remains *for-itself* are non-contingent rights, and such rights are conceptually inconsistent with the application of property law.

The further ambiguity is that the body can be simultaneous *for-itself* and *for-others*. Since our connections or 'intentional ties' with the world include communing and interacting with other embodied subjects, our bodies become 'the medium of social perception' that is a 'medium common to myself and others'.[38] Upon the death of a person, the body is no longer *for-itself* but it may continue to function as the medium of social experience (*for-others*). The relationships that we form with another person, that are mediated through

[38] T Carman, *Merleau-Ponty* (The Routledge Philosophers, Routledge, 2008) 141.

our bodies, may provide a sufficient justification for why we ought to be able to exercise control over the bodies of our loved-ones. These rights of control over bodily material *for-others* also enable preferences orchoices that can only be exercised by a rights-holder. Another person cannot stand in the same position as ourselves with regards to the body of our partner, child, parent, or friend. If we were deprived of the ability to determine how and by whom the body of a loved-one is used and treated, there would be normative consequences (beyond mere allocative consequences) since we would be deprived of a set of preferences and choices that a subsequent possessor of the body cannot exercise.

c. Conceptual Inconsistency

Herein lies the fundamental objection to the application of property rights to all instances of the use or storage of bodily material. On a conceptual level, we know that property is good at dealing with contingent things, but sometimes our control of things is particular to us. Privacy is a useful example. We control our personal information in order to protect a feature of our personality or individuality, and when our control of our personal information is interfered with, the loss suffered is a loss that is particular to, or necessarily associated with, the rights-holder.

Equally, when a parent is deprived of control of a miscarried foetus or the body of a deceased infant, when a widow is unable to continue with assisted reproductive therapy, or when a patient's stored semen thaws, when a directly donated organ is misallocated, the entitlement-holders suffer a loss that is particular to, or necessarily associated with them. In these scenarios the parent, wife, or progenitor are unable to exercise entitlements in bodily material that they have as a parent, as a wife, or as an embodied person. In other words, parents, partners, and patients stand in relation to the bodily material in their capacity as parents, partners, or as physical-subjects. No other person can stand in the same position as the parents, partner, or progenitor with regard to the bodily material. Their rights to the bodily material, similar to the right to bodily integrity and the right to privacy, represent preferences and choices that are necessarily associated with them as the rights-holder.

As we can anticipate, this conceptual inconsistency will have structural implications. The appropriate legal protection for entitlements in bodily material that is *for-itself* or *for-others* requires a responsiveness to preferences and choices that are necessary to the rights-holder. Yet, since a corrective remedial duty and the transferability of the right are structural features that follow from the contingency of the right, property rights represent a structure or rights and duties that is non-responsive to preferences and choices that are necessary to the rights-holder.

d. The Ambiguous Body and the Analogy with Privacy

In our conceptual analysis so far, we have been able to draw upon a similarity between rights in bodily material and property rights since they are both exclusive rights. We have also been able to draw upon similarity between rights in bodily material (that is *for-itself* and *for-others*) and the right to bodily integrity since they protect preferences and choices that are necessarily associated with the rights-holder. However, both items of property and the body are distinct from items of bodily material in one important regard. Property rights and rights of bodily integrity are unambiguous rights. The rights that arise in bodily material are, as we have seen, ambiguous.

The relationship between a person and their body is clear and categorical. As a result, the self-owner has, in all instances, exclusive rights of use and control over his or her body. All other persons are under a duty of non-interference and the duty can only be diffused through the provision of consent. Property is also clear and categorical. The possessor of an item of property has the right to possess the thing that imposes a correlative duty of non-interference, which may only be subject to a superior right to possession. For a duty-bearer, their duty to abstain from interference is a clear and ascertainable duty. In contrast, the relationship between a person and his or her own bodily material is ambiguous; the bodily material may be a mere object (*in-itself*) or may be an expression, extension, or representation of the person's subjectivity (bodily material *for-itself* or *for-others*). The status or value of bodily material is ultimately self-ascribed by the progenitor, and thus—for a duty-bearer—the status of an item of bodily material is unclear and uncertain.

The right to privacy confronts a similar ambiguity. All information that is about or pertains to an identifiable person may be 'personal information' in a broad sense, but it does not follow that the right to privacy amounts to the right to control all information that pertains to the rights-holder. The right to privacy is premised on the value of the rights-holder being able to express themselves selectively, such as relinquishing intimate details in select relationships, reserving some personal details from public judgment and criticism, or constructing a public persona that differs from a person's private individuality. The right to privacy is therefore concerned with the personal information that is reasonably taken or perceived, by the rights-holder, to be information that is intimate, properly reserved from criticism, or pertains to a person's private individuality rather than their public persona. Personal information that is caught within the right to privacy is, in other words, *self-ascribed*.

Hence, in considering the legal structure that ought to apply to the use and storage of bodily material, the right to privacy is a useful analogy. Akin to bodily material (that is *for-itself* or *for-others*) the right to privacy is an exclusionary but non-contingent right. Moreover, similar to the right to use and control bodily material (that is *for-itself* and *for-others*), what is subject to the right to privacy is a matter of self-ascription.

III. Rights, Duties, and Bodily Material

We are now in a position where we can identify the appropriate structure of rights and duties that ought to apply to the use and storage of bodily material. As the earlier discussion has indicated, where rights of use and control arise in bodily material, such rights may not be adequately protected through the application of statutory rights and duties. Given the inadequacy of the governance strategy, the appropriate legal response therefore is to adopt the exclusion strategy. This strategy, of enabling the rights-holder to exclude all others from the object or resource, has a threefold structure that is employed to protect property rights, the right to bodily integrity, and (as we shall see) the right to privacy.

An important complicating factor is that the right to control or use bodily material may enable preferences and choices that can, or cannot, exist independently of the rights-holder. The suggestion has been that, where the rights in the bodily material are contingent to the rights-holder, such rights ought to adopt the remaining structural features of property law. However, where the rights that arise in bodily material are necessarily associated with the rights-holder, a significant departure from the structure of property rights is required. This section will consider how a structural configuration of exclusive but non-contingent rights is possible.

a. The Basis of the Right

Particular structural features ought to follow from the conceptual claim that rights in bodily material are exclusive rights. To start, the basis of the right in bodily material ought to be the relationship with bodily material itself. The implication is that the right to possess the bodily material, at least, ought to be exercisable against an open set of persons. Rights of use and control ought to also be exercisable independent of any interaction with the duty-bearer.

This is not the approach that the law has so far taken. The progenitors in *Moore v Regents of the University of California*,[39] *Washington University v Catalona*,[40] or *Greenberg v Miami Children's Hospital*[41] retained no rights of control over their bodily material *on the basis of their relationship with the material*. The way in which the law has developed is to recognize pre-existing (property) rights in bodily material only after satisfying a further threshold requirement. Namely, a progenitor may obtain pre-existing rights to his or her bodily material upon satisfying the Hegelian test under *Yearworth v North Bristol NHS*.[42] In addition, a third party may obtain pre-existing (property) rights in bodily material subsequent upon satisfying the Lockean test under *Doodeward v Spence*.[43] No person, therefore, has an automatic, immediate, or innate right to his or her own, or another's, bodily material. Under the current legal approach, rights of use and control may arise through an activity or interaction, or where the relationship with the bodily material satisfies a further threshold requirement.

The law's reluctance to recognize 'innate' rights of use and control over bodily material, that are then also exercisable independent of the relationship with the duty-bearer, reflects an ambiguity or complexity that the law does not encounter with items of property nor with the protection of bodily integrity. The ambiguity is that some (but not all) items of bodily material are surplus resources or objects *in-itself*. I have argued that the mere prior embodiment does not justify the right of the progenitor to use or control bodily material. Hence, in *Moore v Regents of the University of California*, *Washington University v Catalona*, and *Greenberg v Miami Children's Hospital* the progenitors had no innate and pre-existing right to control their separated bodily material.

Moreover, it was the combination of pre-existing rights and duties of non-interference under conversion in *Moore v Regents of the University of California* that motivated the Supreme Court of California not to recognize a property in John Moore's spleen cells, since:

> conversion ... imposes liability on all those into whose hands the cells come, whether or not the particular defendant participated in, or knew of, the inadequate disclosures that violated the patient's right to make an informed decision.[44]

[39] *Moore v Regents of the University of California* 793 P 2d 478 (Cal 1990).
[40] *Washington University v Catalona* 437 F Supp 2d 985 (Dist Ct Missouri, 2006).
[41] *Greenberg v Miami Children's Hospital* 264 F Supp 2d 1064.
[42] *Yearworth v North Bristol NHS Trust* (n 10) [45].
[43] *Doodeward v Spence* (1908) 6 *CLR* 406, 414.
[44] *Moore v Regents of the University of California* (n 39) 144.

The concern with pre-existing rights is that they impose duties on the duty-bearer independent of any interaction between the rights-holder and duty-bearer in context where there is ambiguity (for the duty-bearer) as to whether the progenitor retains any rights in the bodily material. There is ambiguity because bodily material may, in some circumstances, be *for-itself* or *for-others*; it may retain a functional unity with the body or it may function as the medium of social experience for others. A progenitor or family member, in such circumstances, may obtain rights of use and control over the bodily material. This problem is further compounded by the fact that where a progenitor or family member does obtain rights of use or control over items of bodily material, they will seldom be in possession of the bodily material.

With regards to property rights and rights of bodily integrity, we can justify the full use of an exclusionary boundary (that *inter alia* imposes duties of non-interference on third parties independent of any action or interaction) because of the ease with which third parties can identify that they are under a duty not to interfere with *the body* or with someone's the *right to possess a thing.* The 'basic inequality' that underlies the exclusionary boundary can be justified with reference to the relative ease with which a duty-bearer can perform duties of non-interference. Hence, when an individual or institution obtains possession of an item of bodily material, their rights in the bodily material ought to be pre-existing rights. Such rights are therefore exercisable against an open set of persons and only subject to any superior rights to the bodily material. Difficulty arises where a person retains the right to use and control bodily material that is not in their possession. In such instances, one structural technique that can be used is to re-adjust the balance between the rights-holder and duty-bearers to limit the correlative duty to circumstances where the duty-bearer knew, or ought to have known, of the entitlements of the rights-holder.

To demonstrate how it is possible to re-adjust the basis of the right to an intermediate position between pre-existing and direct rights, let us consider how duties of confidentiality have been reformulated to protect the right to privacy. The duty of confidentiality concerning the disclosure of personal information is a branch of law that has shifted from requiring a prior interaction or relationship to an intermediate position between pre-existing and direct rights. This traditional formulation of a breach of confidence required that the information 'must have been imparted in circumstances importing an obligation of confidence'.[45] An action for breach of confidence was

[45] *Coco v AN Clark* [1969] RPC 41, 42: 'The information itself... must have a necessary quality about it... It must have been imparted in circumstances importing an obligation of confidence... and a detriment is suffered from the unauthorised use of the information.'

originally only actionable against a set of persons who fell into the relationship of confidence. According to Phillipson, prior to the Human Rights Act 1998, 'it remained the case that there had to be something over and above the quality of the information itself in order to impose an obligation of confidentiality'.[46] In other words, the information not only needed to have a personal or confidential quality about it, but also the circumstances in which the defendant acquired the information had to be circumstances in which a duty of confidence could be imposed on the defendant.

However, since the introduction of the Human Rights Act 1998 the second element of confidentiality has been diluted in an attempt to give greater protection to the right to respect for private and family life under Article 8 of the ECHR. For instance, in *Venables v News Group Newspapers*,[47] the court granted an injunction that prohibited the publication of any information that could identify, or locate, the two youth offenders. The court was explicit in the injunction that the duty not to distribute the information did not depend on the circumstances through which the information was received. In effect, the court 'granted an injunction against the world'[48] and suspended, for this case, the second element of confidential information.

Subsequent case law has also reduced the requirement of a duty of confidence. For instance, the Court of Appeal in *A v B plc* suggested that:

> a duty of confidence will arise whenever the party subject to the duty is in a situation where he either knows or ought to know that the other party can reasonably expect his privacy to be protected.[49]

Although the court in *A v B plc* still maintained the requirement of a duty of confidence, as Phillipson identifies, the 're-working' of the element 'opens up the possibility that such an expectation of privacy could arise simply on the basis of the obviously private nature of the information itself'.[50] Hence, where the quality of the information is particularly personal, this quality in the information will itself impose a duty on persons who acquire the information. The House of Lords in *Campbell v MGN*[51] affirmed the reformulation of the second element. Lord Nichols observed that confidentiality:

> has now firmly shaken off the limiting constraint of the need for an initial confidential relationship. In doing so it has changed its nature … Now the law imposes a

[46] G Phillipson, 'Transforming Breach of Confidence?: Towards a Common Law Right of Privacy under the Human Rights Act' (2003) 66 *MLR* 726, 743.

[47] *Venables and another v News Group Newspapers* [2001] 1 All ER 908.

[48] G Phillipson, 'Transforming Breach of Confidence' (n 46) 744–5.

[49] *A v B plc* [2003] 3 WLR 542, 551.

[50] G Philipson, 'Transforming Breach of Confidence' (n 46) 746.

[51] *Campbell v MGN Ltd* [2004] UKHL 22.

'duty of confidence' whenever a person receives information he knows or ought to know is fairly and reasonably to be regarded as confidential.[52]

This shift in the law of confidential and personal information has made the right to privacy operate more like property rights-based actions insofar as the relationship between the rights-holder and the personal information may be sufficient to ground duties of confidence even where there is no interaction or transaction between the rights-holder and the duty-bearer. That is, provided that the duty-bearer knows or ought to have known of the 'confidential quality' of the information. Hence, under the 'the emerging tort of privacy', the circumstances of how the information is parted, and received, are *relevant to* whether the claimant has a 'reasonable expectation of privacy', but the interaction *is not required* in order for the right to privacy to arise.

These 'latent rights' represent an intermediate position between pre-existing rights (that arise because of the relationship with the thing and are exercisable against an open set of persons) and direct rights (that arise out of a particular activity or interaction involving the duty-bearer). Latent rights are actionable against those who knew, or ought to have known, of the entitlements of the rights-holder. Applying this structural feature to the rights of use and control in bodily material, the right of family members to authorize the use and storage of the body of a deceased person, and the progenitor's right to manage and use his or her own bodily material, are rights actionable against those who knew, or ought to have known, of the entitlements of the family member or progenitor. As we know, the application of the exclusion strategy is not at the expense of the governance strategy. Even where there is a pre-existing right, further rights and duties can be engrafted onto any situation on the basis of the interaction between rights-holder and duty-bearer. For instance, a duty-bearer may obtain possession of bodily material in an interaction that represents an assumption of responsibility for the state of the bodily material or may obtain possession of bodily material that triggers statutory duties with regards to the bodily material.

The key structural innovation suggested here is that rights in bodily material *ought not depend upon* there being an interaction or transaction between the rights-holder and the duty-bearer. Rather, rights may arise because of the relationship with an item of bodily material. In order to account for the ambiguity in bodily material, it is necessary to draw a distinction between possessory and non-possessory rights. The right to possess bodily material

[52] *Campbell v MGN* (n 51) [14].

ought to be a pre-existing right, and as such, is actionable against third parties. Non-possessory interests ought to be latent rights, and therefore be actionable against any person who knew, or ought to have known, of the non-possessory interests, similar to the way the right to privacy is facilitated in the emerging tort of privacy.

b. The Content of the Right

Since exclusive rights protect the relationship between the rights-holder and the thing (the original dimension of the right), as well as the particular activities that the relationship with the thing enables (the derivative dimension of the right), the legal structure that is applied to the use and storage of bodily material ought to include the vindication of both the original and derivative dimensions of the right to possess, control, or use bodily material.

Applying this structural feature to the use and storage of bodily material would vindicate a rights-holder's right to possess, use, or control an item of bodily material without the need to show a tangible or derivative loss. The interference with the right to possess, use, or control bodily material would, itself, be a legal wrong. This development would address the main inadequacy of most non-proprietary solutions to the use and storage of bodily material. As we saw in *Yearworth v North Bristol NHS*, the claimants were unable to show an actionable loss unless the claim was fitted into a property rights-based claim.[53] Moreover, as Remigius Nwabueze has identified, claims for misdirected, misapplied, or maliciously destroyed organs will confront a 'remedial hurdle' in 'proving a contemporaneous loss'[54] unless the legal right adopts some of the structural features of property rights. The remedial hurdle can be cleared because the exclusionary boundary vindicates 'original rights'. The alternative is that a 'derivative loss' is required for a successful action under the governance strategy. Thus, the claimants in *Re: Organ Retention* were required to establish that the physicians' failure to adequately inform

[53] *Yearworth v North Bristol* (n 10) [24]: 'The determination under challenge was to the effect that the decision in *Gregg v Scott* [2005] UKHL 2, [2005] 2 AC 176, compelled a conclusion that, unless the men were to demonstrate a greater than even chance that the lost sperm could have been used in order to achieve conception, then, irrespective of any recovery of their natural fertility, they could not claim in respect of such physical damage to their overall ability to become fathers as was wrought by the so-called personal injury.' *cf Yearworth v North Bristol NHS* (n 10) [25]: 'In order to enable a person to claim in negligence for loss caused to him by reason of loss of or damage to property, he must have had either the legal ownership of or a possessory title to the property concerned at the time when the loss or damage occurred per Lord Brandon in *Leigh and Sillavan Ltd v Aliakmon Shipping Co Ltd* [1986] AC 785 at 809F.'

[54] RN Nwabueze, 'Donated Organs Property Rights and the Remedial Quagmire' (2008) 16 *Med L Rev* 201, 205.

them of the storage and use of their children's organs was not only *wrongful* (breach of duty)[55] but was also *harmful* (caused a derivative loss).[56]

Since the right to privacy is also an exclusive right, duties of confidentiality with regards to personal information have developed to protect the 'original and 'derivative' dimensions of the right. Although in the traditional formulation of the requirements for a breach of confidence in *Coco v Clark*, Justice Megarry stated that 'there must have been an unauthorized use of information *to the detriment* of the party communicating it',[57] Justice Megarry in the same judgment left open whether this requirement was necessary.[58] Lord Keith in *A-G v Guardian (No 2)* then broadened the detriment requirement to recognize, what has been termed here, 'original loss':

> So I would think it a sufficient detriment to the confider that information given in confidence is to be disclosed to persons whom he would prefer not to know of it, even though the disclosure would not be harmful to him in any positive way.[59]

Hence, unauthorized disclosure of confidential information *itself* may be a sufficient infringement to satisfy the third requirement of a breach of confidence. There has been further confirmation that it is unnecessary to show 'detriment' for breaches of medical confidentiality.[60] Hence, the emerging tort of privacy would appear to protect the relationship between the confider and the information without the confider having to show a loss that was derived from the interference with the relationship of control.

Since rights in bodily material represent exclusive rights, it follows that such rights ought to protect the original and derivative dimension of the ownership relationship. Just as with interference with property rights, infringement of the rights of bodily integrity and the unauthorized disclosure of personal information are actionable without the need to show loss or detriment, the interference with rights in bodily material should also be actionable per se.

c. The Content of the Primary Duty

Exclusive rights impose duties to abstain or refrain from action. Property rights impose a duty to refrain from interfering or appropriating items of property; the right to bodily integrity is protected through duties to abstain from non-consensual physical contact; the right to privacy is protected through

[55] *Re: Organ Retention Group Litigation* [2005] QB 506 [170]–[206].

[56] *Re: Organ Retention* (n 55) [259]. [57] *Coco v AN Clark* (n 45) 47 [emphasis added].

[58] *Coco v AN Clark* (n 45) 48; R Arnold, 'The Protection of Confidential Information in the Human Rights Era: Two Aspects' (2007) 2 JIPLP 599.

[59] *A-G v Guardian Newspapers (No 2)* [1988] UKHL 6, [1990] 1 AC 109 (HL) [256].

[60] See *Bluck v Information Commissioner* (2007) 98 BMLR 1, [15].

duties of confidentiality, that prohibit the (intentional or inadvertent) disclosure of personal information.[61] If rights in bodily material are exclusive rights, then such rights also ought to impose duties on the duty-bearer to refrain from interfering with the bodily material.

Yet, our property rights and rights of bodily integrity benefit from the protection provided through complementary duties. Whilst the right-holder's relationship with the object or resource grounds duties on others to refrain from interference, the interaction between the right-holder and the duty-bearer may be sufficient to impose duties of care on the duty-bearer. The interaction amounts to an assumption of responsibility, which is able to explain why the duty-bearer is under an additional set of positive obligations. The legal structure that ought to apply to the use and storage of bodily material is a structure that ought to include both duties to refrain and duties of care. Akin to our relationship with our items of property, bodily integrity, and privacy, the relationship between a rights-holder and an item of bodily material ought to ground duties on an open set of persons to refrain from any voluntary conduct that interferes with the rights-holder's rights of possession, use, or control. An additional duty on the duty-bearer to take reasonable care and the required precautions in order to preserve the rights-holder's rights in the bodily material may also arise ought of any interactions or transactions between the rights-holder and a duty-bearer.

We have been able to identify how the structure of the exclusionary boundary can apply to the use and storage of bodily material. There are three main features to the structure of the exclusionary boundary. First, rights of use of control in bodily material are exercisable against an open set of persons. I have suggested that this feature ought to be adjusted to account for the ambiguity in bodily material. Drawing upon an analogy with the right to privacy, non-possessory rights ought to be exercisable against any person who knew or ought to have known of the non-possessory interest in the bodily material, whereas the right to possess bodily material ought to be exercisable against otherwise anonymous third parties. The second structural feature concerns the content of the right. I have suggested that the rights in bodily material ought to protect both the original and derivative dimension of the ownership relationship. The third feature of the exclusionary boundary is duties to abstain from action. Such duties ought to be imposed on duty-bearers, coupled with duties of care that arise from activities or interactions undertaken by the duty-bearer.

[61] *A-G v Guardian Newspapers (No 2)* (n 59) 34.

d. The Content of the Remedial Duty

Let us now consider the legal structure that follows from the contingency of property rights. I have so far suggested that there is a conceptual inconsistency between property rights and the rights that may arise in bodily material (that is *for-itself* or *for-others*). I explained that two structural features of property law follow from the characterization of entitlements as enabling preferences and choices which are contingent to, or can exist independent of, the rights-holder. The first structural feature that follows from this conceptual characterization is the content of the remedial duty.

As we have seen, property rights impose corrective remedial duties. Under this remedial approach, damages aim to place the rights-holder 'in the same place as if the loss had not been inflicted on him'.[62] To recover a non-pecuniary loss, the rights-holder is required to prove that a non-pecuniary loss, which falls into a recognized category of loss, was foreseeable upon the breach of the primary duty by the duty-bearer. This remedial approach otherwise presumes that correcting the misallocation of entitlements will restore the preferences and choices that the entitlements enabled, unless a further loss can be proven by the claimant.

The way in which the law remedies interference with personal rights is to redistribute benefits and burdens in light of the merit of the parties. This reflects the inability of corrective damages to remedy interference with preference and choices that are necessarily associated with the rights-holder. Hence, with torts that 'concern an individual's personality' distributive remedial duties presume that a personal loss has been suffered and the remedial duty includes 'damages without being required to prove actual emotional harm'.[63]

Given that entitlements in bodily material (that is *for-itself* or *for-others*) arise because the bodily material either retains a functional unity with the body or remains the medium of social experience, the loss suffered from interference with these entitlements will be almost invariably non-pecuniary. Under corrective remedial duties, claimants carry the burden to prove that the non-pecuniary loss fits within a recognized category of loss and prove the foreseeability of the non-pecuniary loss.[64] This burden was discharged by the claimants in *Yearworth v North Bristol NHS*[65] but provided an additional

[62] NE Palmer and E McKendrick (eds), *Interests in Goods* (2nd edn, Lloyd's of London Press, London 1998) 548.

[63] N Witzleb, 'Monetary Remedies for Breach of Confidence in Privacy Cases' (2007) 27 LS 430, 446.

[64] *Yearworth v North Bristol* (n 10) [54].

[65] *Yearworth v North Bristol* (n 10) [10]–[12].

remedial hurdle for the claims in *Re: Organ Retention* that some claimants were unable to clear.[66] In contrast, if distributive remedial duties were applied in response to the interference, it will be presumed that the interference with rights in the bodily material (that is *for-itself* or *for-others*) will cause some form of personal loss, mental distress or anguish, and general damages would be awarded. The concern here is that the appropriate remedial response to the infringement of non-contingent rights is a distributive approach and the application of corrective remedial duties therefore represents an inappropriate remedial response.

Duties of confidentiality with regards to personal information provide a useful example of how a branch of law has developed to award 'general damages' as an ordinary remedial response to interference with non-contingent rights or interests. It is difficult for monetary damages to correct or restore the type of loss suffered from the unauthorized disclosure of personal information. If only pecuniary losses were recoverable from a breach of confidence, the protection of the right to control the material itself (as opposed to the right not to suffer loss from the unauthorized disclosure) would not be adequately protected. As Justice Morland observed in *Cornelius v de Taranto*, the protection of the right of respect for private and family life would be 'hollow' if only nominal damages were awarded in lieu of the claimant being able to prove a pecuniary loss.[67] Hence, Justice Morland allowed general damages for emotional distress.[68] The award of general damages for injury to feelings was affirmed by the House of Lords in *Campbell v MGN*.[69] Following the reasoning in *Cornelius v de Taranto*, the High Court has held that 'where a breach causes injury to feelings, this court has power to award general damages'.[70] The suggestion here is that general damages are available under the emerging tort of privacy (without having to prove a particular category of non-pecuniary loss).

Ultimately, the award of general damages under actions for breach of confidentiality will depend on the 'juridical basis of the cause of action'.[71] Confidentiality, insofar as it governs the disclosure of information, governs a wide variety of contexts where someone confides information. Where the context is commercial, equity learns from the remedial response taken in

[66] *Re: Organ Retention* (n 9) [259].

[67] *Cornelius v de Taranto* [2001] EMLR 12 (QB) [66].

[68] *Cornelius v de Taranto* (n 67).

[69] *Campbell v MGN* (n 51).

[70] *Archer v Williams* [2003] EWHC 1670 (QB) (Justice Jackson) [76]: '[W]here a breach causes injury to feelings, this court has power to award general damages.'

[71] N Witzleb, 'Monetary Remedies for Breach of Confidence' (n 63) 549: '[T]he juridical basis of the cause of action can also assume relevance for the availability of the remedy.'

contract and property law, and where the context concerns privacy, equity draws upon the damages awarded for defamation.[72]

Where the value of the bodily material stems from preferences and choices that can exist independently of the rights-holder (bodily material that is *in-itself*), the aims of corrective justice are applicable. The appropriate remedial response of the law is, therefore, corrective. The relevant juridical basis of the remedial duty will be same as the remedial duties imposed under property law. However, where the value of the bodily material is based upon preferences and choices that cannot exist independently of the rights-holder, the aims of redistributive justice are more applicable. In order to identify the 'juridical basis' for remedying losses with regards to bodily material (that is *for-itself* or *for-others*), the appropriate analogies engage remedial principles that protect preferences and choices that cannot exist independently of the rights-holder, such as, the right to bodily integrity or the right to privacy. In these instances, the law ought to assume that a personal loss follows from the infringement of personal rights in bodily material.

Although distributive justice may be primarily concerned with the compensatory aims of awarding damages in proportion to the loss suffered by the wronged-party, a large range of considerations may become relevant because the distribution of benefits and burdens is determined by reference to the 'merit' of the parties. Since a collection of considerations can shape the relative merits of the parties involved in the dispute, the criteria for allocating benefits and burdens represent an opportunity to advance a particular set of a large range of public policies. For example, Larry Palmer argues that, in the context of biobanks, Guido Calabresi and Douglas Melamed's structure of 'liability rules'[73] enable the courts to exercise a degree of social control by not only forcing research and clinical enterprises to account for their practices but also by limiting the extent individuals can dictate terms to these socially important enterprises.[74] By awarding damages that are not necessarily determined by inter-exchange between the parties Palmer argues that the law can 'attempt to optimize the benefits and risks of knowledge distribution' in biobanking[75] since (to borrow Nwabueze's description):

> [L]iability rules are superior to property rules in dealing with issues raised by emergent biobanks because they bring into focus the relevant ethical

[72] N Witzleb, 'Monetary Remedies for Breach of Confidence' (n 63) 448.

[73] G Calabresi and AD Melamed, 'Property Rules, Liability Rules and Inalienability: One View of the Cathedral' (1972) 85 *Harv L Rev* 1089, 1092.

[74] LI Palmer, 'Should Liability Play a Role in Social Control of Biobanks?' (2005) 33 *J Law Med Ethics* 70.

[75] LI Palmer, 'Should Liability Play a Role in Social Control of Biobanks' (n 74) 70.

issues, such as a person's right to control the actions of professionals and their organisation.[76]

Charlotte Harrison also advocates the use of 'liability' or 'governance' rules for the protection of entitlements in separated bodily material in an approach that avoids the commodification of the person and would 'draw from models of publicly mediated compensation in other areas of law'.[77] Under this approach recommended here, the law would employ a distributive justice approach to primarily address the need to compensate the non-pecuniary losses that follow from interference with personal interests. Yet, a range of public policy factors, as Harrison and Palmer suggest, may also guide the allocation benefits and burdens under the recommended approach. These factors and considerations are not adequately captured under the remedial response of property law.

There is, therefore, a structural limit to the appropriate legal response that property rights can provide for the interference with rights in bodily material. This structural problem reflects a deeper conceptual unease. Since property law is concerned with contingent rights, it is 'blind' or unresponsive to the preferences and choices that are particular to the entitlement-holder. In other words, property law misses, by-passes, or ignores the preferences and choices that are particular to, or necessarily associated with, the entitlement-holder. Hence, rights of use and control over items of bodily material (that is for-itself or for-others) ought to be protected by distributive remedial duties.

The suggestion here has been that the infringement of rights in bodily material that is *in-itself* ought to be remedied through the imposition of corrective remedial duties. However, such corrective duties provide an inadequate remedial response to the infringement of rights in bodily material that is *for-itself* or *for-others*. In order for the law to be responsive to preferences and choices that are particular to the rights-holder, rights in bodily material that is *for-itself* or *for-others* ought to be vindicated by distributive remedial duties.

e. The Transferability of the Right

The right to bodily integrity is an inalienable or non-transferable right. The right to privacy, insofar as it protects an ambiguous subject-matter, is more

[76] RN Nwabueze, 'Remedial Quagmire' (n 54) 202.

[77] CH Harrison, 'Neither Moore nor the Market: Alternative Models for Compensating Contributors of Human Tissue' (2002) 28 *Am JL & Med* 77, 94.

complex. Although a number of people may possess the personal information that pertains to a particular person, only that particular person to whom the information pertains to is the privacy rights-holder. Although personal information (the subject-matter) is transferable, the right to privacy (the legally protected relationship with the subject matter) is non-transferable. In comparison, property rights are transferable rights. Property rights may be transferable upon agreement between the transferor and transferee where the parties themselves determine the content of the exchange. Alternatively, the transfer of property rights may also be subject to liability or governance rules that determine the content and circumstances of the transaction, in order to account for the wider normative consequences of the transfer for the object or resource.

Since a progenitor retains the right to control and use bodily material that retains a functional unity with the body, such rights are necessarily associated with the rights-holder (and co-participants in treatment). It is because of this necessary association with the rights-holder (including the reversionary-holder) that such rights are non-transferable. Similarly, the rights of family members to control how the body of a deceased loved-one is treated are rights that are explained in terms of the body of the deceased continuing to be the medium of social experience. Such rights are rights that that ought to be limited to those particular individuals. The reason why those rights are non-transferable is because the transfer of the right dislocates the *exercise* of the right from the *justification* of the right. If these rights *were* transferred to another person, the reasons justifying the exercise of the entitlement would no longer be applicable to the recipient of the right. The transfer of rights in bodily material that is *for-itself* and the bodily material is *for-others* removes the justification for why the entitlements ought to be legally protected.

This prescription is at odds with recent judicial developments where property law has been employed to facilitate the transfer of entitlements in bodily material to the partner of the progenitor of reproductive material. Once bodily material is conceived of as an item of property there are few procedural or substantive limits on the right to transfer. The application of succession legislation has enabled the transfer of semen to the widows of the deceased progenitors in *Hecht v Superior Court of Los Angeles County*[78] and *Bazely v Wesley Monash IVF*.[79] Moreover, as the reasoning of *Re: Edwards* demonstrates, claims to use the bodily material removed from an unconscious or deceased person may succeed through the application of property law even

[78] *Hecht v Superior Court of Los Angeles County* (1993) 20 Cal Rptr 2d 275.
[79] *Bazely v Wesley Monash IVF* [2010] QSC 118.

when the consent of the progenitor is required under statute for the use of the bodily material.[80]

For body material that is *in-itself*, there are no normative consequences from the change in rights-holder. This is because any such rights of use or control in bodily material (that is *in-itself*) are only contingently associated with the rights-holder. Such rights in bodily material, therefore, ought to be (presumptively) transferable rights. However, this presumption of transferability may be overturned. The transfer of bodily material may nonetheless have wider normative consequences, and therefore attract the application of liability rules. Statutory provisions often govern how bodily material is transferred to another person in response to concerns over the authenticity of consent, the welfare of the transferor, and the appropriate content of the exchange. It is because of the extent of these statutory limits and controls, that the transferability of bodily material may, in some legal systems, be entirely determined by governance or liability rules.

I have argued here that rights in bodily material, that are justified either because the bodily material remains the medium of a person's engagement with the world (bodily material *for-itself*) or continues to be the medium of social experience (bodily material that is *for-others*) ought to give rise to non-transferable rights. The change in entitlement-holder removes the justificatory basis for the exercise of the entitlement. Hence, a further structural problem with conflating the distinction between contingent rights and non-contingent rights is that it enables the transferability of rights that ought not to be inalienable. In other words, it treats an ownership relationship that has no equivalence as an ownership relation that has value equivalence.

We therefore have two structural problems with the application of property law to bodily material (that is *for-itself* or *for-others*). Because property law is oriented towards preferences and choices that can exist independently of the rights-holder, property law has a narrow conception of how the mistreatment of objects or resources can be remedied as well as a permissive approach as to transferability of objects and resources. As I have argued here, bodily material (that is *for-itself* and *for-others*) is conceptually distinct from items of property, and therefore ought to attract a different legal structure. The implication throughout this structural analysis has been that property rights represent the same set of rights and duties that ought to apply to bodily material that is *in-itself*, and the emerging tort of privacy represents the same set of rights and duties that ought to apply to bodily material that is *for-itself* and *for-others*.

[80] Assisted Reproductive Technology Act 2007 (NSW), s 21.

IV. Structural Configurations and the Way Further Forward

In this chapter I have formulated a recommendation as to how the law should develop. The recommendation has been in terms of structural configurations that ought to apply to the use and storage of bodily material. The particular structural configuration that ought to apply will depend on whether the rights in the bodily material are contingently or non-contingently associated with the rights as well as whether the rights-holder is in possession of the bodily material.

a. Structural Configurations

Allow me to first audit these structural configurations. I have argued that all rights that arise in bodily material ought to be protected by the structure of the exclusionary boundary. In terms of *the content of the right*, the law ought to protect both the original and derivative dimension of the relationship between the rights-holder and the bodily material. This removes the need for the rights-holder to identify a loss subsequent to any interference with the bodily material. In terms of *the content of the primary duty*, the right to possess, control, or use bodily material ought to be protected through a duty of non-interference with the bodily material. Additional duties of care and precaution may also be imposed on the duty-bearer where the duty-bearer has engaged in activities and interaction that represent an assumption of responsibility by the duty-bearer.

I have also suggested that the exclusion boundary ought to be amenable to narrowing the category of persons against whom the right is actionable. In terms of *the basis of the right*, where the rights-holder is in possession of the bodily material, their rights ought to be actionable against an open set of persons. Where the rights-holder is not in possession, his or her rights ought to actionable against those who knew, or ought to have known, of the rights-holder's non-possession rights in the bodily material. This adjustment reflects an ambiguity the bodily material; that from the point of view of the duty-bearer it is not clear whether an item of bodily material is *for-itself*, *for-others*, or *in-itself*.

How the remaining structural features apply to the use and storage of bodily material depends on a distinction between bodily material that is *in-itself* and bodily material that is *for-itself* (or *for-others*). Where the bodily material is *in-itself*, the appropriate legal structure includes the imposition of corrective remedial duties and the right to transfer the bodily material ought to arise (although the right to transfer may be limited by liability or governance rules). Alternatively, where the bodily material is *for-itself* or *for-others*, the appropriate remedial duties ought to be guided by the principle of distributive justice and any rights in the bodily material ought to be non-transferable.

These structural configurations are *akin to* existing legal doctrines. The structure of property law is a structure of exclusive and contingent rights that are orientated towards the possession of an object or resource. The structure of the emerging common law right to privacy is a structure of exclusive and non-contingent rights that is able to govern an ambiguous subject matter. Property rights and the right to privacy are, therefore, *relevant structural analogies* for the appropriate legal status of bodily material. By identifying the appropriate structural configurations for the law that governs the use and storage of bodily material, this chapter has fallen short of fully identifying the appropriate *legal status* of bodily material. The question of the legal status of bodily material has become, at this stage in the overall analysis, a question of labelling. The structural analogies provide the way forward. In practical terms, there are two main ways to develop the law in adherence to these structural analogies.

b. A New Property

It may be that all rights in bodily material obtain the status or label of 'property rights'. In which case the structure of such property rights require adjustment. Such adjustments are needed to address both the ambiguity in the body and the non-contingent relationships that form between a rights-holder and an item of bodily material (that is *for-itself* or *for-others*).

In terms of the ambiguity of the body, since exclusive rights of use and control do not innately vest in the rights-holder (a progenitor, family member, or healthcare institution), some test or threshold is required to determine when such property rights arise in bodily material in consideration of when duty-bearers can be expected to perform the correlative duties of non-interference. The common law has so far developed a rudimentary means of differentiation through affirming, whilst also identifying exceptions to, the 'no property rule'. Under the common law, there are no innate rights in bodily material unless either the thresholds in Lockean[81] or Hegelian[82] tests are satisfied or the duty to inter the body of the deceased arises.[83]

[81] *Doodeward v Spence* (n 43) 414; *R v Kelly* (n 13) 750; *Dobson and another v North Tyneside Health Authority and another* [1996] 4 All ER 474.

[82] *Yearworth v North Bristol NHS* (n 10) [45].

[83] *Dobson and another v North Tyneside Health* (n 81) 478, quoting *Clerk and Lindsell on Torts* (17th edn, Sweet & Maxwell, London 1995) 653 [13–50]: 'the executors or administrators or other

Two structural changes are then required in order for property law to account for the preferences and choices that cannot exist independently of the rights-holder. First, the property rights that a progenitor (or family member) exercises in bodily material ought to be non-transferable. Viewing the circumstances in *Bazely v Wesley Monash IVF*, *Hecht v Superior Court of Los Angeles County*, and perhaps in *ex parte Blood*, and *Re: Edwards*,[84] as the exercise of reversionary entitlements can relieve the pressure on the law to permit the transfer of rights in bodily material. In addition, where the property rights of a progenitor or family member are infringed, the remedial hurdles that obstruct the award of 'general damages' ought to be removed. Rather than the burden being placed on the claimant to prove a particular category of loss that was foreseeable upon breach of the primary duty,[85] a breach of a primary duty with regards to bodily material (that is *for-itself* or *for-others*) ought to be presumed to cause a general category of non-pecuniary loss. These are both significant modifications to the structure of property rights that are necessary if property law is to adequately govern the use and storage of bodily material.

c. A New Confidentiality

Alternatively, two distinct legal categories can be created. An entitlement in bodily material that is *in-itself* may become a property right. Where the law permits a healthcare institution to possess bodily material, their set of entitlements in the material may be protected by the property rights-based actions. An entitlement in bodily material that is *for-itself* (or *for-others*), in comparison, can be governed by a structure of rights and duties that is akin to the duties of confidentiality with regards to personal information.

The current law on when entitlements in bodily material vest in progenitors, family members, or healthcare institutions can be starting points for the task of differentiating bodily material that is *in-itself* from bodily material that is *for-itself* or *for-others*. Hence, the Hegelian basis in *Yearworth v North Bristol NHS* or the statutory rights of progenitors may ground duties of confidentiality with regards to bodily material that is *for-itself*. In addition,

persons charged by the law with the duty of interring the body have a right to the custody and possession of it until it is properly buried'; see also *Re: Gray* [2000] QSC 390, 390.

[84] *Bazely v Wesley Monash IVF* (n 79); *Hecht v Superior Court of Los Angeles County* (n 78); *R v Human Fertilisation and Embryology Authority, ex parte Blood* [1997] 2 All ER 687; *Jocelyn Edwards; Re the Estate of the Late Mark Edwards* [2011] NSWSC 478.

[85] *RK v Oldham NHS Trust* [2003] Lloyd's Rep Med 1 (QB) [20]; R Mulheron, 'Rewriting the Requirement for a "Recognised Psychiatric Injury" in Negligence Claims' (2012) 32 *Oxford J Legal Stud* 77. See also D Nolan, 'New Forms of Damage in Negligence' (2007) 70 *MLR* 59, 61:

the right to possess as the corollary of the duty to inter recognized *Dobson v North Tyneside Health Authority*, the reversionary interests of co-participants in treatment, and the statutory rights of spouses and family members, may also ground duties of confidentiality with regards to bodily material that is *for-others*. The Lockean basis in *Doodeward v Spence* and the series of statutory provisions that enable healthcare institutions to possess, use, control, transfer, and profit from bodily material that is *in-itself* may be the point upon which healthcare institutions obtain property rights in bodily material. It follows that, in most instances where progenitors or family members retain entitlements in bodily material, healthcare institutions who possess the items of bodily material may obtain property rights in bodily material that are nonetheless subject to duties of confidentiality owed to progenitors or family members.

This dualist approach requires a new legal category: duties of confidentiality with regards to bodily material. This represents a novel doctrinal development. The extent of doctrinal change is arguably in equal proportion to the recognition of property rights in material that has been subject to a 'no property rule' for over three centuries of legal history. In either case, significant doctrinal developments are required to fill the void left by the 'no property rule'.

These two recommendations as to how the law ought to develop represent different ways of categorizing the same legal structure. All rights in bodily material are exclusive rights, the set of persons against whom the right is actionable will depend on whether the right-holder is in possession of the bodily material, and remedial duties and the right to transfer will vary depending on the relationship between the rights-holder and the item of bodily material. At this point (and only at this point) in the overall analysis, it becomes 'a question of labelling'[86] whether these rights in bodily material all represent 'property rights'.

V. Chapter Summary

To conclude, let us return to *Evans v Amicus Healthcare* and *AW v CC*. Mr Johnson, by providing semen to form embryos, was exercising his reproductive autonomy, a reproductive preference, and his right to respect for his

[86] *cf* R Rao, 'Property, Privacy and the Human Body' (2000) 80 *BUL Rev* 359, 460: 'If society permits a part of the body to be separated from the person and alienated to others or seized by the state, that part of the body constitutes "property", regardless of the legal label attached to it.'

private and family life. His decision to withhold his consent for the use of the embryos engages the same components of his subjectivity, and engages his personal rights. Whether his interest in non-procreation is sufficient to defeat Ms Evans 'all-or-nothing' chance to conceive a child can be doubted. In either case, the interests involved are conceptually distinct from a property interest. I have formulated here a way in which the law can govern the possession, use, and control of bodily material without having to reduce all the interests that arise in bodily material to property rights.

In contrast, when AW provided his semen to form embryos, AW was transferring a surplus resource. It was a gifting of property; a relinquishment of any interests in the object or resource material in the context of a private agreement. This act of relinquishment is common place insofar as we alienate a range of bodily material from ourselves because the bodily material is not engaged in a current or potential task. I have formulated here a way in which the law can recognize that some items of bodily material are capable of being items of property.

The aim of this chapter has been to identify the particular set of rights and duties that ought to apply to the use and storage of bodily material. In order to get a sense of how configurations of rights and duties differ, we have considered how rights in bodily material conceptually and structurally align with property rights, the right to bodily integrity, and the right to privacy. Through this analysis we have identified some of the strengths and weaknesses of property law as the response to disputes over the use and storage of bodily material.

Our rights in our items of property, our bodies, and our personal information all represent an interest in an open set of activities that is best protected through the power to exclude others. I have argued that the entitlements that arise in bodily material are conceptually similar. Insofar as bodily material may continue to be functionally unified with the body, may continue as the medium of social experience, or may be used in pursuit of a series of socially valuable aims, the entitlements that arise in bodily material represent an interest in purposively dealing with a *thing* rather than an interest in a particular *activity*. The current governance strategy of allocating statutory rights and duties between participants to an activity therefore struggles to capture the fullness of the entitlements in bodily material.

The way in which property rights differ from the right to bodily integrity and the right to privacy can be understood in terms of whether the preferences and choices that underlie the right can exist independently of the rights-holder. When we apply this distinction to the rights that arise in bodily material, we encounter two conceptually distinct ownership relationships: a contingent relationship between a rights-holder and bodily material that is *in-itself* and a non-contingent relationship between a rights-holder and bodily material that is *for-itself* or *for-others*. I have argued here that there is a conceptual inconsistency between property rights and rights in bodily material (that is *for-itself* or *for-others*). Furthermore, there is ambiguity in bodily material since it is capable of being material that is *in-itself*, *for-itself*, or *for-others*.

The solution, I have suggested, lies in identifying the appropriate configuration of rights and duties. Simply put, rights in bodily material that is *in-itself* ought to adopt the structure of property rights. Rights in bodily material that is *for-itself* or *for-others* ought to adopt the structure of the right to privacy. This dualist approach has been informed and motivated by an underlying philosophical contention: that the world is made of things and that some things, even things that are physically separate from a person, may nonetheless be interwoven with a person's being or subjectivity. Although property law is well equipped to govern the use of objects, it is nonetheless poorly equipped to govern the objects that constitute our subjectivity.

Conclusion

By way of conclusion, allow me to briefly map the analysis from the preceding discussion as against the way in which the law on the use and storage of bodily material has developed. We will then be able to fully appreciate the current state of the law and how the recommendations formulated in this book provide a way for the law to develop.

I. The No Property Rule

The common law is premised upon the basic rule that there is no property in the body or its parts. There are two main implications of the 'no property rule'. The first is that the progenitor of bodily material does not have an innate right to possess, control, use, or profit from his or her own bodily material. Hence, in *Williams v Williams*, the deceased was unable to make directions as to the disposal of his body through his will,[1] in *Greenberg v Miami* the donors of blood and tissue were unable to control the use of the isolated gene,[2] in *Washington University v Catalona* the donors of tissue were unable to determine who was able to conduct research on their bodily material,[3] and in *Moore v Regents of the University of California* the progenitor was unable to share in the profits from a cell-line derived from his wrongfully obtained spleen cells.[4]

This implication of the 'no property rule' is a negation of the otherwise intuitive principle that a person's prior embodiment of the bodily material justifies why a person ought to retain entitlements in his or her bodily

[1] *Williams v Williams* (1882) 20 Ch D 659.
[2] *Greenberg v Miami Children's Hospital*, 1064, 264 F Supp 2d 1066.
[3] *Washington University v Catalona* 437 F Supp 2d 985 (Dist Ct Missouri, 2006).
[4] *Moore v Regents of the University of California* 793 P 2d 478 (Cal 1990).

material. The problem with the prior embodiment principle is that it does not explain why prior acquisition or possession is a past act or event that can, in the present, impose on others duties to refrain from using the resource. I have suggested that there is no reason to presume that someone's prior relationship with bodily material is connected to the current assertion of rights in the bodily material. Rather, something further is required in order to show why the progenitor's prior relationship with the bodily material is sufficient to explain why he or she ought to be able to—in the present—control or use the bodily material.

The second implication is that where an individual or institution is able to exercise rights of use or control over bodily material, the structure of such rights are shaped by a governance strategy that allocates rights and duties between participants in an activity or interaction. For instance, in *Re: Organ Retention* the parents of the deceased children had to either identify a breach of statutory duty or establish that: the physicians owed a duty of care to the parents of the deceased children on the bases of the factual circumstances of the interaction;[5] the parents suffered a loss;[6] and the loss was a foreseeable loss flowing from physician's breach of the duty of care.[7]

Instances arose where the allocation of rights and duties between participants in an activity provided an inadequate form of legal protection for entitlements in bodily material. In *Yearworth v North Bristol NHS*, although the patients had statutory rights to control and use their semen, such statutory provisions were able to respond to the infringement of the patients' right to use their semen caused by inadequate storage of the bodily material.[8] Moreover, in *R v Kelly*, although the Royal College of Surgeons may have had a statutory right to possess the preserved body parts, the statutory right could not form the basis of claim against the artist who removed and disposed of the college's specimens.[9]

The inadequacies of the governance strategy provided the impetus for exceptions to the 'no property rule' to arise. The benefit of conceptualizing entitlements in bodily material as property rights is the structure of protection provided by the 'exclusionary boundary'.

5 *Re: Organ Retention Group Litigation* [2005] QB 506 [170]–[206].
6 *Re: Organ Retention* (n 5) [568]. 7 *Re: Organ Retention* (n 5) [259].
8 *Yearworth v North Bristol NHS Trust* [2009] EWCA Civ 37 [44].
9 *R v Kelly* [1998] 3 All ER 741, 750.

II. The Lockean Exception

The first structural inadequacy of the governance strategy is that it is only able to impose duties on participants to particular activities or interactions. Property rights that employ an exclusion strategy are able to impose duties on an open set of persons. Hence, the Lockean exception formulated in *Doodeward v Spence*,[10] and applied in *R v Kelly*, ushered items of bodily material behind an exclusionary boundary that imposed duties on an open set of persons not to interfere with, or appropriate, the items of property.

The exception itself, which requires the 'lawful exercise of work or skill' that results in the corpse acquiring 'some attributes differentiating it from a mere corpse',[11] can be understood in two different ways. It is possible to interpret the Lockean exception as providing a pre-social basis for entitlements in the bodily material. Such an explanation contends that when a person applies work or skill on an object, his or her labour is mixed with the object, and to remove the object from the labourer's control is to deprive the person of ownership of his or her labour. However, this account of ownership, based upon the single causal link between actions of the person and the attributes of the object, is over-inclusive. Creating a new attribute in the object is, by itself, insufficient to explain why the object is brought within the sphere of the subject instead of the work or skill being 'dissipated' into the external world.

A more promising interpretation of the 'work or skill exception' to the no property rule is that the exception represents a social basis for entitlements in bodily material. According to this explanation, the 'application of work or skill' is an activity that the law seeks to incentivize or encourage. For instance, the preservation of body parts for medical or scientific examination is socially valuable, and in order to encourage the activity, the law protects the labourer's' entitlement to possess, use, or even profit from the body material.

III. The Hegelian Exception

The second structural inadequacy of the governance strategy concerns the content of the right. Since the focal point of the law under the governance strategy is the activity or interaction that the rights-holder and duty-bearer is engaged with, the content of the right is limited to the 'derivative' dimension of the right. Hence, in *Yearworth v North Bristol NHS*, the patients' claim

10 *Doodeward v Spence* (1908) 6 CLR 406.

11 *Doodeward v Spence* (n 10) 414.

for negligence causing injury or distress failed because they could not prove, on the balance of probabilities, that damage to the semen interfered with their ability to conceive a child.[12] This motivated the second exception to the no property rule. It is because, the Court of Appeal reasoned, the claimants alone generated and ejaculated the sperm, where the sole object of their ejaculation was that it might later be used for their benefit, 'the men had ownership of the sperm for the purposes of their present claims ... namely in relation to its future use'.[13] Their relationship with their bodily material provided a possessory interest, the interference with which was actionable per se.

The explanation for why the progenitors owned their bodily material aligns with a Hegelian account of ownership. According to Hegel, 'a person must translate his freedom into an external sphere in order to exist as Ideal'.[14] The way in which an element of a person's subjectivity is expressed in an object is through a 'nexus of dual relations between the will and object, object and will'.[15] Where an object acquires an attribute or characteristic 'which can only be explained in terms of the working will',[16] and where an object 'affects the will' by enabling a certain state of affairs that the individual intends, desires, or 'wills' to exist that would otherwise not exist without the object, the 'dual nexus' is formed. This 'dual nexus' is a sound basis upon which to justify the ownership of bodily material.

IV. Beyond Hegel and Beyond *Yearworth*

However, the reasoning in *Yearworth v North Bristol NHS* is only able to sustain a particular scope of claims. Since the 'dual nexus' requires an attribute 'which can only be explained in terms of the working will', the reasoning cannot support claims of ownership over bodily material removed from an unconscious or deceased person. The progenitor, in such circumstances, does not have 'a working will'. Hence, in the subsequent application of the reasoning in *Yearworth v North Bristol NHS* to posthumous cases such as *Re: Edwards*, courts have formulated a wider principle, that a person other than the progenitor may obtain the right to possess or use bodily material where bodily material is extracted 'for their purposes'.[17] Such reasoning

[12] *Yearworth v North Bristol NHS Trust* [2009] EWCA Civ 37 [24].

[13] *Yearworth v North Bristol NHS* (n 12) [45(f)].

[14] AW Wood (ed), GWF Hegel, *Elements of the Philosophy of Right* (HB Nisbet (tr)) (Cambridge University Press, Cambridge 1991) [41].

[15] J Waldron, *The Right to Private Property* (Clarendon Press, Oxford 1988) 370.

[16] J Waldron, *The Right to Private Property* (n 15) 369.

[17] *Jocelyn Edwards; Re the Estate of the Late Mark Edwards* [2011] NSWSC 478, [91].

pre-supposes that the purposes for which the bodily material is removed are purposes that the claimant is legally entitled to pursue, and is otherwise justified in doing so. The problem with the posthumous applications of the reasoning in *Yearworth v North Bristol NHS* is that it is circular to suggest that a person can have bodily material removed from another person, and then point to the fact it was removed 'for their purposes' in order to validate the purposes for which they intend to use the bodily material.

It is nonetheless possible to broaden the reasoning in *Yearworth v North Bristol NHS*. The Hegelian 'dual nexus' is a sound justificatory basis as it is able to identify a set of instances where the connection between, or integration of, a person's body and his or her subjectivity is sustained. I have argued that where bodily material retains a functional unity with the body, the integration of a person's body and the person's subjectivity is sustained. In such instances, where the bodily material is unified with the body in response to a task (bodily material that is *for-itself*) or continues to function as the medium of social experience (bodily material that is *for-others*), entitlements to possess, use, and control the bodily material ought to arise.

The reasoning in *Yearworth v North Bristol NHS* is also unable to sustain claims to the right to profit from the use, relinquishment, or transfer of bodily material. More broadly, any pre-social account of ownership is unable to justify the right to profit. This is because the considerations that form a pre-social account of ownership (the attributes, characteristics, and moral properties of a person) are of a different order of considerations than the considerations that form the content of the right to profit (the economic and social conditions that determine the monetary value of the good). The content of the right to profit is 'distributionally determined'[18] whilst the attributes and characteristics of the person are 'distributionally blind'.[19] If we are to recognize the right to profit from the use, relinquishment, or transfer of bodily material, then such a right can only be justified with reference to the state of affairs that the exercise of the right is likely to produce. Moreover, we ought not to readily assume that the availability of a financial benefit will increase the supply of bodily material, whilst also being aware of the risk of value denigration that arises when financial benefits are offered to motivate the decision to relinquish or transfer bodily material.

[18] J Christman, 'Distributive Justice and the Complex Structure of Ownership' (1994) 23 *Philos Public Aff* 225, 231.

[19] JW Harris, *Property and Justice* (Oxford University Press, Oxford 1996) 246.

V. The Current State of the Law

By formulating exceptions to (rather than departures from) the 'no property rule', the law has so far preserved the presumption that prior embodiment is not itself sufficient to obtain rights of control or use. Instead, a claimant must satisfy a further threshold. An important feature of the current law is the way in which it is caught midway between a general 'no property rule' with emerging and expanding exceptions to the rule. This reflects an ambiguity that is unfamiliar to property law. The ambiguity is that parts of our bodies and our bodily material may be mere surplus resources or may be the site or location of our subjectivity. For instance, the semen that was provided as a 'unconditional gift' to form the embryos in *AW v CC*[20] is materially the same as, but also fundamentally distinct from, the semen provided to form the embryos in *Evans v Amicus Healthcare*.[21] It is therefore necessary to differentiate bodily material that is a surplus material resource from bodily material that retains a connection to the person. This task arises because of the unique status of the body, and by extension bodily material, that is both a material and physical object and the location of our subjectivity. I have suggested that whether an item of bodily material is a surplus resource or the location of our subjectivity is matter of self-ascription; we determine for ourselves what parts of our body or bodily material are engaged in our tasks, plans, projects, preferences, and desires (or any other component of subjectivity).

The common law right to privacy is more familiar with this ambiguity. All information that pertains to an identifiable person may be 'personal information' in a broad sense, but it does not follow that the right to privacy amounts to the right to control all information that pertains to the rights-holder. The right to privacy is concerned with the personal information that is intimate, properly reserved from criticism, or pertains to a person's private individuality rather than their public persona. Similar to our relationship with our bodies, the personal information that is governed by the right to privacy is *self-ascribed*. The common law then protects the right to privacy 'whenever a person receives information he knows or ought to know is fairly and reasonably to be regarded as confidential'.[22]

The main feature of the current state of the law that this discussion has aimed to isolate is how the application of property law has introduced two

[20] *CC v AW*, 2005 ABQB 290 [20].

[21] *Evans v Amicus Healthcare and others* [2004] EWCA Civ 727; *Evans v United Kingdom* (2008) 46 EHRR 34 (Grand Chamber).

[22] *Campbell v MGN Ltd* [2004] UKHL 22, [14].

conceptual features of the law that govern the use and storage of bodily material. It has introduced exclusive rights that protect an open-ended set of activities through the exclusion of others from the object or resource. This has been a welcome development in the law, as it provides a structure of legal protection that is more suited to the use and storage of bodily material. This structure includes rights that are exercisable against an open set of persons, that protect the original dimension of the right and impose primary duties of non-interference.

However, in the same way that the Hegelian account of property creates a paradox in attempting to show a sufficiently close whilst sufficiently contingent connection between a person and an object or resource, the recognition of property rights in bodily material following *Yearworth v North Bristol NHS* creates a conceptual inconsistency. I have argued that a person ought to retain entitlements in bodily material where, as in *Yearworth v North Bristol NHS*, the material retains a functional unity with the body. In these instances, the entitlements in the bodily material represent a set of preferences and choices that cannot exist independently of the rights-holder. Property rights, however, are contingent rights; they protect preferences and choices that can exist independently of the rights-holder. This conceptual inconsistency has structural implications. Applying the structural features of contingent rights to rights that are necessarily associated with the rights-holder results in a legal structure that is too restrictive in terms of recognizing non-pecuniary loss whilst too permissive in terms of the transferability of the right.

The discussion throughout the previous chapters has been motivated by this inconsistency between a series of important ownership relationships between a person and bodily material (that is *for-itself* or *for-others*) and the conceptual and structural features of property law. We can avoid the dichotomy if we are able to uncouple the exclusionary boundary from the contingency of property law.

VI. Maintaining a Distinction in the Law

The recommendations developed here represent an attempt to resist the coupling of the exclusionary boundary with contingency rights in the law that applies to the use and storage of bodily material. Both the right to bodily integrity and the right to privacy are exclusive but non-contingent rights; they are able to provide the structural protection of the exclusionary boundary whilst also protecting preferences and choices that cannot exist independently of the rights-holder. The right to privacy also grapples with the same ambiguity of meaning and value that arises in the context of the use and storage of bodily material. That is, there are some instances where bodily material retains a connection with the progenitor as well as many instances where the use of bodily material is equivalent to the use of any other material resource. Hence, I have argued that the way forward for the law is to operate through two structural configurations: the structure of property rights-based actions and the structure of the emerging tort of privacy.

This dualist recommendation follows first from the conceptual cleaving of ownership from property; we accept that individuals and institutions ought to be able to own bodily material without having to accept that every ownership relationship is a property relationship. This dualist recommendation also follows from a particular understanding of the body and bodily material. That is, the understanding that our bodies are our point of view on the world 'on which' we 'can take no point of view'.[23] As such, the structures and capacities of the body 'can never manifest themselves as contingent and arbitrary for us'.[24] To the extent that bodily material may continue to be the medium through which we engage in the world, and engage with others, our relationship with bodily material also must be a necessary and constitutive (rather than contingent and arbitrary) relationship. To then protect this relationship through a branch of law that is oriented towards preferences and choices that are contingent to the rights-holder is a mistake that conflates the distinction between things that have value in their own right and things that have substitutional value. To align these different values 'along a single metric' does 'violence to our considered judgment about how these goods are best characterized'.[25]

There is no longer the jurisdictional distinction between the civil and ecclesiastical cognizance that, perhaps by way of a historical accident, gave rise to the rule that 'the burial of the *Cadaver*... is nullius in bonis [the property of no one], and belongs to Ecclesiastical cognizance'.[26] I have argued

[23] SK Toombs, 'What Does it Mean to Be Somebody?: Phenomenological Reflections and Ethical Quandaries' in MJ Cherry (ed) *Persons and Their Bodies: Rights, Responsibilities, Relationships* (Kluwer Academic Publishers, Dordrecht 1999) 79.

[24] T Carman, *Merleau-Ponty* (The Routledge Philosophers, Routledge, 2008) 82–3.

[25] Cass R Sunstein, 'Incommensurability and Kinds of Valuation: Some Applications in Law' in R Chang (ed) *Incommensurability, Incomparability and Practical Reason* (Harvard University Press, Cambridge 1997) 234, 238.

[26] E Coke, *The Third Part of the Institutes of the Laws of England: Concerning High Treason, and Other Pleas of the Crown, and Criminal Causes* (Flesher, Lee & Pakeman, London: 1644) 203; PDG Skegg, 'Human Corpses, Medical Specimens and the Law of Property' (1975) 4 *Anglo-Am L Rev* 412; PDG Skegg, 'The "No Property" Rule and Rights Relating to Dead Bodies' (1997) 5 *Tort L Rev* 222, 222–9.

that we ought to nonetheless maintain an underlying distiction that has been articulated in different ways throughout this discussion. That is, a distinction between the external relationship that we have with the objects that we own and the existential relationship that we have with our bodies; between things that have substitutional value and things that have value in their own right; between legal rights that can exist independently of us as the rights-holder and legal rights that are necessarily associated with us as the rights-holder. The contention here has been that, when it comes to the possession, use, and control of bodily material, we ought to maintain a distinction in the law between our ownership of contingent and substitutional things and our respect for each others' physical subjecivity and being.

Bibliography

BOOKS AND BOOK CHAPTERS

Anderson, E, *Value in Ethics and Economics* (Harvard University Press, Cambridge MA 1993).

Benn, S, 'Privacy, Freedom and Respect for Persons' in F Schoeman (ed) *Philosophical Dimensions of Privacy: An Anthology* (Cambridge University Press, Cambridge 1984).

Bennett, B, *Health Law's Kaleidoscope: Health Law Rights in a Global Age* (Ashgate, Aldershot 2008) 95.

Bollier, D, *Silent Theft: The Private Plunder of Our Common Wealth* (Routledge, New York 2002).

Carman, T, *Merleau-Ponty* (The Routledge Philosophers, Routledge, 2008).

Carter, A, *The Philosophical Foundations of Property Rights* (Harvester Wheatsheaf, London 1989).

Cohen, GA, 'Self-Ownership, Freedom and Equality' in N Bunnin and J Yu (eds) *The Blackwell Dictionary of Western Philosophy* (Blackwell Publishing, Malden MA 2004).

——— *Self-Ownership, Freedom and Equality* (Cambridge University Press, Cambridge 1995).

Coke, E, *The Third Part of the Institutes of the Laws of England: Concerning High Treason, and Other Pleas of the Crown, and Criminal Causes* (Flesher, Lee & Pakeman, London 1644).

Coleman, JL, *The Practice of Principle* (Oxford University Press, Oxford 2001).

Critchley, S, *Ethics-Politics-Subjectivity* (Verso, London 1999).

Davis, MJ and N Naffine, *Are Persons Property? Legal Debates about Property and Personality* (Ashgate, Aldershot 2001).

Donner, W, 'A Millian Perspective on the Relationship between Persons and Their Bodies.' in MJ Cherry (ed) *Persons and Their Bodies: Rights, Responsibilities, Relationships* (Kluwer Academic Publishers, Dordrecht 1999).

Douglas, S, 'Property Rights in Human Biological Material' in I Goold and others (eds), *Persons, Parts and Property: How Should We Regulate Human Tissue in the 21st Century?* (Hart Publishing, Oxford 2014).

Dworkin, G, 'Markets and Morals: The Case for Organ Sales' in G Dworkin (ed) *Morality, Harm, and the Law* (Westview, Boulder, Colo 1994).

Endicott, T, 'The Infant in the Snow' in T Endicott, J Getzler, and E Peel (eds) *Properties of Law* (Oxford University Press, Oxford 2006).

Fabre, C, *Whose Body Is It Anyway? Justice and the Integrity of the Person* (Oxford University Press, Oxford 2006).

Ferguson, KE, *The Man Question: Vision of Subjectivity in Feminist Theory* (University of California Press, Berkley 1993).
Foster, C, *Human Dignity in Bioethics and Law.* (Hart Publishing, Oxford 2011).
Fried, C, *An Anatomy of Values: Problems of Personal and Social* (Harvard University Press, Cambridge 1970).
Gardner, J, 'Obligations and Outcomes in the Law of Torts' in P Cane and J Gardner (eds) *Relating to Responsibility: Essays for Tony Honoré on His Eightieth Birthday* (Hart Publishing, Oxford 2001).
Green, S, 'Rights and Wrongs: An Introduction to the Wrongful Interference Actions' in D Nolan and A Robertson (eds), *Rights and Private Law* (Hart Publishing, Oxford 2012).
Griffin, J, *On Human Rights* (Oxford University Press, Oxford 2008).
Hardcastle, R, *Law and the Human Body: Property Rights, Ownership and Control* (Hart Publishing, Oxford 2009).
Harris, JW, *Property and Justice* (Clarendon Press, Oxford 1996).
———*Wonderwoman and Superman: The Ethics of Human Biotechnology* (Oxford University Press, Oxford 1992).
Hegel, GWF, HB Nisbet (tr), AW Wood (ed), *Elements of the Philosophy of Right* (Cambridge University Press, Cambridge 1991).
Herring, J, *Medical Law and Ethics* (4th edn, Oxford University Press, Oxford 2011).
———'Why We Need a Statute Regime to Regulate Bodily Material' in I Goold et al, *Persons, Parts and Property: How Should We Regulate Human Tissue in the 21st Century?* (Hart Publishing, Oxford 2014).
Hill, J, 'The Proprietary Character of Possession' in E Cooke (ed), *Modern Studies in Property Law: Volume 1* (Hart Publishing, Portland 2001).
Honoré, AM, 'Ownership' in Honoré, AM (ed), *Making Law Bind: Essays Legal and Philosophical* (Clarendon Press, Oxford 1961).
Husserl, E, M Biemel (ed), *Ideen zu einer reinen Phänomenologie und phänomenologischen Philosophie. Zweites Buch: Phänomenologische Untersuchungen zur Konstitution. Husserliana IV* (Martinus Nijhoff, The Hague 1952).
Kant, I, *Lectures on Ethics* (Harper & Row, New York 1963).
———MJ Gregor (ed) *The Groundwork of the Metaphysics of Morals* (Cambridge University Press, Cambridge 1996).
Kimbrell, A, *The Human Body Shop: The Engineering and Marketing of Life* (Harpercollins, San Francisco 1993).
Macpherson, CB, 'The Meaning of Property' in CB Macpherson (ed) *Property: Mainstream and Critical Positions* (Basil Blackwell, Oxford 1978).
McFarlane, B, *The Structure of Property Law* (Hart Publishing, Oxford 2008).
McMillan, J, 'Identity, Self, and Dementia' in Huges et al (eds) *Dementia: Mind, Meaning and the Person* (Oxford University Press, Oxford 2006).
Merleau-Ponty, M, C Smith (tr), *Phenomenology of Perception* (Routledge, Oxford 2003).
Mill, JS, *Utilitarianism* in J Robson (ed) *The Collected Works of John Stuart Mill, Volume 10—Essays on Ethics, Religion and Society* (Toronto University Press, Toronto 1969).

———*On Liberty* in J Robson (ed) *The Collected Works of John Stuart Mill, Volume 18—Essays on Politics and Society Part I* (Toronto University Press, Toronto 1985).

Munzer, SR, *A Theory of Property* (Cambridge University Press, Cambridge 1990).

Nozick, R, *Anarchy, State, and Utopia* (Basic Books, New York 1974).

Palmer, NE and E McKendrick (eds), *Interests in Goods* (2nd edn, Lloyd's of London Press, 1998).

Parfit, D, *Reasons and Persons*. (Clarendon Press, Oxford 1984).

Powers, TM, 'The Integrity of Body: Kantian Moral Constraints on the Physical Self' in MJ Cherry (ed) *Persons and Their Bodies: Rights, Responsibilities, Relationships* (Kluwer Academic Publishers, Dordrecht 1999).

Price, D, 'Property, Harm and the Corpse' in B Brooks-Gordon (ed) *Death Rites and Rights* (Hart Publishing, Oxford 2007).

Priest, S, *Merleau-Ponty* (Arguments of the Philosophers Series, Routledge, London 1998).

Radin, MJ, *Contested Commodities* (Harvard University Press, Cambridge MA 1996).

Rawls, J, *A Theory of Justice* (Harvard University Press, Cambridge MA 1971).

Raz, J, *Practical Reason and Norms* (Oxford University Press, Oxford 1999).

Sartre, JP, H Barnes (tr) *Being and Nothingness: A Phenomenological Essay on Ontology* (Philosophical Library, New York 1956).

Shildrick, M, *Leaky Bodies and Boundaries: Feminism, Postmodernism and (Bio)ethics* (Routledge, London 1997).

Stevens, R, *Torts and Rights* (Oxford University Press, Oxford 2007).

Sunstein, CR, 'Incommensurability and Kinds of Valuation: Some Applications in Law' in R Chang (ed) *Incommensurability, Incomparability and Practical Reason* (Harvard University Press, Cambridge 1997).

Swadling, W, 'Unjust Delivery' in A Burrows and A Rodger (eds) *Mapping the Law: Essays in Memory of Peter Birks* (Oxford University Press, Oxford 2006).

Taylor, JS, *Stakes and Kidneys: Why Markets in Human Body Parts Are Morally Imperative* (Ashgate, Aldershot 2005).

Tasioulas, J, 'On the Nature of Human Rights' in G Ernst and J-C Heilinger (eds) *The Philosophy of Human Rights: Contemporary Controversies* (De Gruyter, Berlin 2012).

Titmuss, RM, *The Gift Relationship: From Human Blood to Social Policy* (Allen & Unwin, London 1970).

Toombs, SK, 'What Does it Mean to Be Somebody?: Phenomenological Reflections and Ethical Quandaries', in MJ Cherry (ed) *Persons and Their Bodies: Rights, Responsibilities, Relationships* (Kluwer Academic Publishers, Dordrecht 1999).

Underkuffler, LS, *The Idea of Property: Its Meaning and Power* (Oxford University Press, Oxford 2003).

Waldron, J, *The Right to Private Property* (Clarendon Press, Oxford 1988).

Walzer, M, *Spheres of Justice: A Defence of Pluralism and Equality* (Basic Books, New York 1983).

Weinrib, E, *The Idea of Private Law* (Harvard University Press, Cambridge MA 1995).

Wittgenstein, L, GEM Anscombe (tr) *Philosophical Investigations* (Basil Blackwell, Oxford 1968).
Zaner, RM, *The Context of Self: A Phenomenological Inquiry Using Medicine as a Clue* (Ohio University Press, Athens 1981).

ARTICLES

Almeling, R, 'Why Do You Want to Be a Donor?': Gender and the Production of Altruism in Egg and Sperm Donation' (2006) 25 *New Genet Soc* 143.
Arnold, R, 'The Protection of Confidential Information in the Human Rights Era: Two Aspects' (2007) 2 *JIPLP* 599.
Atkins, K, 'Autonomy and the Subjective Character of Experience' (2000) 17 *J Appl Philos* 71.
Boulier, W, 'Sperm, Spleen and Other Valuables: The Need to Recognise Property Rights in Human Body Parts' (1995) 23 *Hofstra L Rev* 693.
Calabresi, G and AD Melamed, 'Property Rules, Liability Rules and Inalienability: One View of the Cathedral' (1972) 85 *Harv L Rev* 1089.
Cameron, JS and R Hoffenberg, 'The Ethics of Organ Transplantation Reconsidered: Paid Organ Donation and the Use of Executed Prisoners as Donors' (1999) 55 *Kidney Int* 727
Cane, P, 'Causing Conversion' (2002) 118 *LQR* 544.
Carman, T, 'The Body in Husserl and Merleau-Ponty' (1999) 27(2) *Phil Topics* 205, 206.
Chadwick, RF, 'The Market for Body Parts: Kant and Duties to Oneself' (1989) 6 *J Appl Philos* 129.
Christman, J, 'Distributive Justice and the Complex Structure of Ownership' (1994) 23 *Philos Public Aff* 225.
———'Self-Ownership, Equality and the Structure of Property Rights' (1991) 19 *Polit Theory* 28.
———'Can Ownership Be Justified by Natural Rights?' (1986) 15 *Philos Public Aff* 156.
Cohen, G, 'The Price of Everything, The Value of Nothing: Reframing the Commodification Debate' (2003) 117 *Harv L Rev* 689.
Coleman, JL 'Corrective Justice and Property Rights' (1994) 11(2) *Soc Phil & Pol'y* 124.
Cronin, A and J Harris, 'Authorisation, altruism and compulsion in the organ donation debate' (2010) 36 *J Med Ethics* 627.
Daniels, KR, R Curson, and GM Lewis, 'Semen Donor Recruitment: A Study of Donors in Two Clinics' (1996) 11 *Hum Reprod* 746.
Deci, EL, 'Effects of Externally Mediated Rewards on Intrinsic Motivation (1971) 18 *J Pers Soc Psychol* 105.
Douglas, S, 'The Nature of Conversion' (2009) 68 *CLJ* 198.
Dworkin, G, 'Is More Choice Better Than Less?' (1982) 7 *Midwest Stud Philos* 47, 61.
Fabre, C, 'Reply to Wilkinson' (2008) 14 *Res Publica* 137.

Ford, M, '*Evans v United Kingdom*: What Implications for the Jurisprudence of Pregnancy?' (2008) 8 *HRL Rev* 171.
Frey, BS and R Jegen, 'Motivation Crowding Theory' (2001) 15 *J Econ Surv* 589.
Gardner, J, 'What is Tort Law For? Part 1. The Place of Corrective Justice' (2011) 30 *Law & Phil* 1.
George, A, 'The Difficulty in Defining Property' (2005) 25 *Oxford J Legal Stud* 793.
Getzler, J, 'Theories of Property and Economic Development' (1996) 26 *J Interdiscip Hist* 639.
Gillespie, TW and CH Hillyer, 'Blood Donors and Factors Impacting the Blood Donation Decision (2002) 16 *Transfus Med Rev* 115.
Goold, I, 'Sounds Suspiciously like Property Treatment: Does Human Tissue Fit within the Common Law Concept of Property?' (2005) 7 *UTS L Rev* 62.
Green, S, 'The Subject Matter of Conversion' [2010] *JBL* 218.
———'Understanding the Wrongful Interference Actions' (2010) 74 *Conv* 15.
Greenawalt, K, 'Privacy and Its Legal Protections' (1974) 2 *Hastings Centre Stud* 45.
Hansson, O and B Bjorkman, 'Bodily Rights and Property Rights' (2006) 32 *J Med Ethics* 209.
Harmon, SHE and GT Laurie, '*Yearworth v North Bristol NHS Trust:* Property, Principles, Precedents and Paradigms' (2010) 69 *CLJ* 486.
Harrington, M, MR Sweeney, K Bailie and others, 'What Would Encourage Blood Donation in Ireland?'(2007) 92 *Vox Sanguinis* 361.
Harris, J, 'Law and Regulation of Retained Organs: The Ethical Issues' (2002) 22 *LS* 527, 548.
Harrison, CH, 'Neither Moore nor the Market: Alternative Models for Compensating Contributors of Human Tissue' (2002) 28 *Am JL & Med* 77.
Hart, HLA, 'Are There Any Natural Rights?' (1955) 64 *Phil Rev* 175–91.
Herring, J and P-L Chau, 'My Body, Your Body, Our Bodies' (2007) 15 *Med L Rev* 34.
Hughes, PM, 'Exploitation, Autonomy, and the Case for Organ Sales' (1998) 12 *Int J Appl Philos* 89.
Kass, LR, 'Organs for Sale? Propriety, Property, and the Price of Progress' (1992) 107 *Pub Int* 65.
Katz, L, 'Exclusion and Exclusivity in Property Law' (2008) 58 *UTLJ* 280.
Laurie, GT, 'Challenging Medical-Legal Norms: The Role of Autonomy, Confidentiality, and Privacy in Protecting Individual and Family Group Rights in Genetic Information' (2001) 22 *J Legal Med* 1.
Liddell, K and A Hall, 'Beyond Bristol and Alder Hey: The Future of Regulation of Human Tissue' (2005) 13 *Med L Rev* 170.
Madden Dempsey, M, 'Victimless Conduct and the Volenti Maxim: How Consent Works' (2013) 7 *Crim L & Phil* 11.
Mason JK and GT Laurie, 'Consent or Property? Dealing with Body Parts in the Shadow of Bristol and Alder Hey' (2001) 64 *Mod L Rev* 710.
Mellström, C and M Johannesson, 'Crowding Out in Blood Donation: Was Titmuss Right?' (2008) 6 *J Eur Econ Assoc* 845.

Merrill, T and HE Smith, 'The Morality of Property' (2007) 45 *Wm & Mary L Rev* 1849.

Merrill, T and HE Smith, 'What Happened to Property in Law and Economics' (2001) 111 *Yale LJ* 257.

Mill, JS, *On Liberty* in J Robson (ed) *The Collected Works of John Stuart Mill, Volume 18—Essays on Politics and Society Part I* (Toronto University Press, Toronto 1985) 261.

Mulheron, R, 'Rewriting the Requirement for a "Recognised Psychiatric Injury" in Negligence Claims' (2012) 32 *Oxford J Legal Stud* 77.

Munzer, SE, 'Uneasy Case against Property Rights in Body Parts' (1994) 11 *Soc Phil & Pol'y* 259, 286

Naffine, N, 'The Structure of Self-Ownership: Or the Self Possessed Man and the Woman Possessed' (1998) 25 *JL & Soc'y* 193.

Nedelsky, J, 'Law, Boundaries and the Bounded Self' (1990) 30 *Representations* 162.

Nolan, D, 'New Forms of Damage in Negligence' (2007) 70 *MLR* 59.

Nwabueze, RN, 'Donated Organs Property Rights and the Remedial Quagmire' (2008) 16 *Med L Rev* 201.

Palmer, LI, 'Should Liability Play a Role in Social Control of Biobanks?' (2005) 33 *J Law Med Ethics* 70.

Partridge, E, 'Posthumous Interests and Posthumous Respect' (1981) 91 *Ethics* 243.

Phillipson, G, 'Transforming Breach of Confidence?: Towards a Common Law Right of Privacy under the Human Rights Act' (2003) 66 *MLR* 726.

Quigley, M, 'Property and the Body: Applying Honoré' (2007) 33 *J Med Ethics* 631.

———'Property in Human Biomaterials—Separating Persons and Things?' (2012) 32 *Oxford J Legal Stud* 659.

Radin, MJ, 'What, If Anything, Is Wrong with Baby Selling?' (1995) 26 *PAC LJ* 135.

———'Property and Personhood' (1982) 34 *Stan L Rev* 957.

———'Market Inalienabilty' (1987) 100 *Harv L Rev* 1849.

———'The Colin Raugh Thomas O'Fallan Memorial Lecture on Reconsidering Personhood' (1995) *Or L Rev* 423.

———'Compensation and Commensurability' (1993) 43 *Duke L J* 56.

Rao, R, 'Property, Privacy and the Human Body' (2000) 80 *BU L Rev* 359, 460.

Reich, C, 'The New Property' (1964) 73 *Yale LJ* 733.

Rippon, S, 'Imposing Options on People in Poverty: The Harm of a Live Donor Organ Market' (2014) 40 *J Med Ethics* 145.

Savulescu, J, 'Is the Sale of Body Parts Wrong?' (2003) 29 *J Med Ethics* 138.

Schreiber, GB, KS Sclumpf, SA Glynn, and others, 'Convenience, the Bane of Our Existence, and Other Barriers to Donating' (2006) 46 *Transfusion* 545.

Siminoff, L, MB Mercer, G Graham, and others, 'The Reasons Families Donate Organs for Transplantation' (2007) 62 *J Trauma* 969.

Simmons, J, 'Original-Acquisition Justifications of Private Property' (1994) 11 *Soc Phil & Pol'y* 63.

Skegg, PDG, 'Human Corpses, Medical Specimens and the Law of Property' (1975) 4 *Anglo-Am L Rev* 412.

Skegg, PDG, 'The "No Property" Rule and Rights Relating to Dead Bodies' (1997) 5 *Tort L Rev* 222, 222–9.

Skene, L, 'Proprietary Interests in Human Bodily Material: Yearworth, Recent Australian Cases on Stored Semen and Their Implications' (2012) 20 *Med L Rev* 227.

Smith, HE, 'Modularity and Morality in the Law of Torts' (2011) 4 *J Tort L* 1

———'Property and Property Rules' (2004) 79 *NYU L Rev* 1719.

Spital, A and JS Taylor, 'Routine Recovery of Cadaveric Organs for Transplantation: Consistent, Fair, and Life-Saving' (2007) 2 *Clin J Am Soc Nephrol* 300.

Sque, M, T Long, S Payne, and others, 'Why Relatives Do Not Donate Organs for Transplants: "Sacrifice" or "Gift of Life"?' (2008) 61 *J Adv Nurs* 134.

Trindade, FA, 'Intentional Torts: Some Thoughts on Assault and Battery' (1982) 2 *Oxford J Legal Stud* 211.

Waldron, J, 'Two Worries about Mixing One's Labour' (1983) 33 *Phil Q* 39.

Weinrib, EJ, 'Corrective Justice in a Nutshell' (2002) 52 *UTLJ* 349.

Wilkinson, TM, 'The Confiscation and Sale of Organs' (2007) 13 *Res Publica* 327, 328.

Witzleb, N, 'Monetary Remedies for Breach of Confidence in Privacy Cases' (2007) 27 *LS* 430.

REPORTS, REGULATIONS, AND COMMAND PAPERS

Bristol Royal Infirmary Inquiry, 'The Report of the Public Inquiry into Children's Heart Surgery at the Bristol Royal Infirmary 1984–1995: Learning from Bristol' (Cm 5207, 2001) Annex C: 'Interim Report: Removal and Retention of Human Material' [33].

Department of Health, *Annual Report of the Chief Medical Officer* (2001).

Hansard HC col 115 (28 June 2004) (Ms Winterton).

Human Fertilisation and Embryology Authority, *Code of Practice* (8th edn, London 2013).

Nuffield Council on Bioethics, *Human Bodies: Donation for Medicine and Research* (2011).

Index

The manufacturer's authorised representative in the EU for product safety is Oxford University Press España S.A. of El Parque Empresarial San Fernando de Henares, Avenida de Castilla, 2 - 28830 Madrid (www.oup.es/en or product.safety@oup.com). OUP España S.A. also acts as importer into Spain of products made by the manufacturer.
Printed and bound by CPI Group (UK) Ltd, Croydon, CR0 4YY
14/09/2026
02216304-0004